AF553712

CELL BIOTECHNOLOGY

CELL BIOTECHNOLOGY

By

Dr. Rajeev Tyagi

Department of Zoology
M.M. (P.G.) College
Modi Nagar (U.P.)

DISCOVERY PUBLISHING HOUSE
NEW DELHI-110002

First Published-2005

ISBN 81-7141-960-7

Published by

DISCOVERY PUBLISHING HOUSE
4831/24, Ansari Road, Prahlad Street,
Darya Ganj, New Delhi-110002 (India)
Phone: 23279245 • Fax: 91-11-23253475
E-mail:dphtemp@indiatimes.com

Printed at:
Amit Enterprises, Delhi

Preface

"Cell Biotechnology" is an effervescent science, a new wave of science, a verdant field of research. It is at once intensely technical and intensely commercial. This branch of biotechnology powerfully diverted the attention of biologists, chemists, engineers and indeed all manner of technically minded professional and lay people. Its commercial relevance inevitably attracts the curiosity not only of people who just wish to understand what is going on around them but also of those whose business it is to take stock of the world of industry, and to perceive patterns of development and the nature and sizes of present and future markets, in order to decide where investment should or should not be made. Understanding in any depth the impact that biotechnology will increasingly have on human affairs, or making commercial decisions about technical investments, must be more than a little difficult for anyone whose education and training does not include the science and engineering on which the technology is based.

Cell biotechnology today is undergoing a resurgence in a wide range of applications and it was to this end that the concept of writing present title is originated. The present title is primarily designed for under- and post-graduate students of all Indian Universities, and it is assumed that these students already have basic knowledge of Cell Biology, Botany, Genetics and Microbiology.

To make the work more comprehensive and informative, the author has consulted many authoritative books, research journals, abstracts, monographs etc. He is grateful to all those great scholars whose work are cited or substantially reproduced.

There can be no claim to originality except in the manner of treatment and much of the information has been obtained from the books and scientific journals available in the different libraries.

The author expresses his thanks to his friends and colleagues whose continue inspirations have initiated him to bring out this book.

The author is painfully aware of the shortcomings, errors and misprints that have crept in, and shall be greateful to receive suggestion for improvement of the next edition from all the readers.

The author expresses his gratitude to Mr. Wasan and staff of M/s Discovery Publishing House for their whole hearted co-operation in the publication of this book.

Author

Contents

1

Morphology of Cell

The cell is a unit of biological activity delimited by a semipermeable membrane and capable of self-reproduction in a suitable non-living medium. The body of all living organisms (bacteria, blue green algae, plants and animals) except viruses has cellular organization and may contain one or many cells. The organisms with only one cell in their body are called acellular organisms (e.g., bacteria, blue green algae, some algae, protozoan, etc.). The organisms having many cells in their body are called multicellular organisms (e.g., most plants and animals). Any cellular organism may contain only one type of cell Prokaryotic cells, Eukaryotic cells. It was *Robert Hooke* who first of all in 1665 observed under a microscope certain honey comb like structures in cork, and he applied the term 'cell' (L., *cella;* small room) for the same. Previously it was believed that there is only one component of the cell which is cell wall. But in 1831, one more peculiar structure was observed by *Robert Brown.* He gave the name 'nucleus'. Later on the 1835 the word 'Sarcode' was proposed for the jelly like material present inside the cell by *Dujardin.* In 1840 *Purkinje* replaced the word sarcode by Protoplasm which is now used universally.

Prokaryotic Cells

The prokaryotic (Gr., *pro*-primitive, *Karyon*-nucleus) cells are the most primitive cells from morphological point of view. They occur in bacteria and blue green algae. Prokaryotes are small, single cells organism, usually less than a micrometer (abbreviated mm; 1000 μm = 1 millimeter, abbreviated mm) are generally not longer than 3 μm. Their structural organization is simple than that of the eukaryotes.

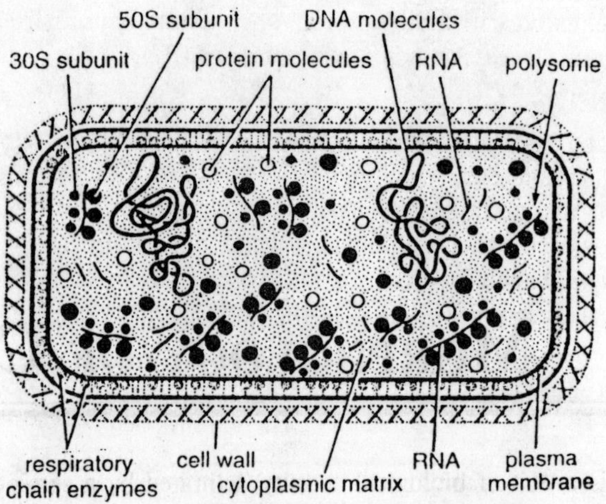

Fig. 1.1. A prokaryotic cell of Escherichia coli.

They do not have a distinct nucleus, and they are lacking in many membranous structures found in the more complex cell types. The hereditary information is contained in a single circular structure that lacks many of the molecular components of the eukaryotic hereditary structure called the chromosome. In addition, the prokaryotes seem to lack any kind of special apparatus for cell division that is why they divide amitotically or by fission.

The prokaryotes are characterized by:

1. the basic protein—histones, which are one of the most important constituents of chromosomes of eukaryotic cells, are absent in prokaryotic chromosomes.
2. the genetic material is located on a single chromosome which consists of a circular double strand of DNA.
3. the absence of a membrane around the nuclear material.
4. the cell wall is non-cellulosic, being formed of carbohydrates and amino acids.
5. the absence of clearly defined membrane-limited organelles like mitochondria, chloroplast, Golgi complex and lysosomes.
6. the absence of nucleolus and mitotic apparatus.
7. the plasma membrane which lies below the cells wall is produced into the cytoplasm and acts as the mitochondrial membrane carrying respiratory enzymes, and
8. the cytoplasm neither exhibits streaming nor the amoeboid movement.

The prokaryotes include bacteria, blue green algae, spirochaete, *Mycoplasma* or pleuropneumonia like organism (PPLO).

Bacterial Cell

The bacteria are microscopic, cellular, achlorophillous, asexually reproducing prokaryotes which occur in fresh and salt water, in soil and in plants and animal. They lead either a saprophytic or parasitic mode of existence. The saprophytic species of bacteria are of great economic significance for man. Some parasitic species of bacteria are pathogenic (disease producing) to plants, animals and man.

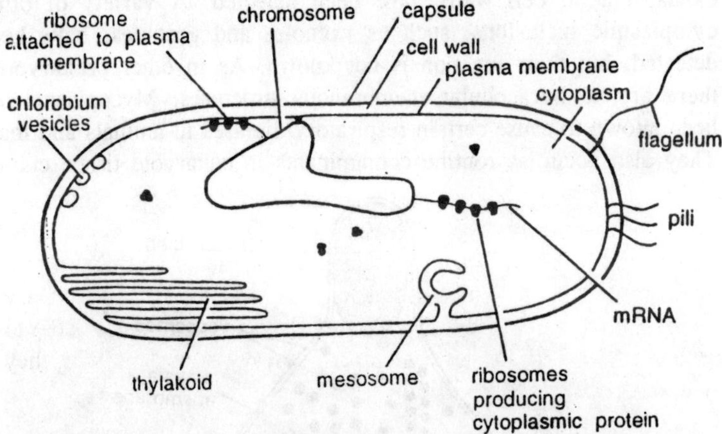

Fig. 1.2. A diagram of structures seen in the prokaryotic (bacterial) cells.

Bacteria are very small having an average diameter of 1.25 μ. The smallest bacterium is *Dialister pneumosintes* (0.15 to 0.3 μ) and largest bacterium is *Spirillum volutans* (13-15 μ) in length. One of the best studied bacteria is *Escherichia coli*. It is about 2 μm long and 1mm wide and contains several thousand different kinds of molecules. Structurally, *E. coli* is rod like and has a typical cell wall and underlying cell membrane. The membrane contains the special regions called *mesosomes*, which apparently have a role in cellular respiration and cell division. There may be one or several "nucleoid" areas, which contain a single, circular molecule of hereditary information (DNA). The rest of the bacterial cytoplasm is full of structures called ribosomes, which differ slightly from the ribosomes of eukaryotic cells. These structures are the sites of protein synthesis. In addition, recent studies have revealed the presence of small circular DNA molecules in the cytoplasm called *plasmids*. These structures are very important in genetic engineering.

Mycoplasma

The *Mycoplasma* (previously called PPLO, pleuropneumonia like organisms) are small (0.2 to 0.3 μm) and more simply organized than bacteria. The hereditary material (DNA) is contained in a non-membrane-bound, nucleus like region, and it may exist either as strands or as a circular molecule as in the bacteria. However, there is only one tenth the amount of genetic material in Mycoplasma as there is in bacteria. Similarly, the Mycoplasma contains 1/50 to 1/100 the number of bacterial ribosomes per cell. A delimiting cell membrane exists, but no cell walls have been detected. A variety of other cytoplasmic inclusions, such as vacuoles and granules, have been detected, but their function is not known. As in other prokaryotes, there are no intracellular membranous structures. Mycoplasma has been known to cause certain respiratory diseases in animals and man. They also occur as routine contaminants in eukaryote tissue culture

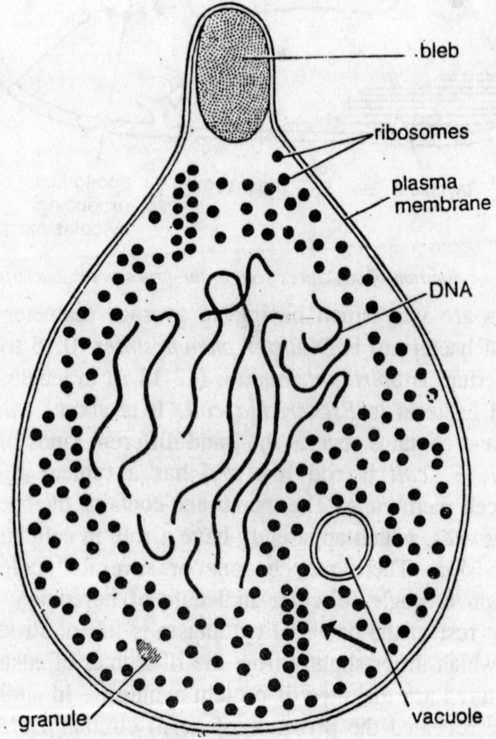

Fig. 1.3. A schematic diagram of typical PPLO cell.

and currently are of much concern to researchers who use these culture procedures in their studies.

Blue-Green Algae

The blue-green algae are the third major group of prokaryotic organisms. The cellular organization is more comples than that of the other prokaryotes, primarily because of the existence of photosynthetic machinery. The blue-green algae contain chlorophyll as well as other more unusual pigments called *phycobilins*. The chlorophylls are bound to membranes in flattened rows called lamellae, which give the algal cytoplasm a rather organized and structured appearance. The phycobilins appear to be located in granular structures called *cyanosomes,* which apparently are attached to the outer lamellar membrane surface. Like the other prokaryotes, the blue-green algae do not have a membrane-bound nucleus or other membrane-delimited organelles. The genetic material exists as free strands of DNA, and the ribosomes are free in the cytoplasm as in the bacteria and *Mycoplasma.* There appears to be other cytoplasmic granules of a protein and phosphate nature, and the entire cell is surrounded by a cell wall with an underlying plasma membrane.

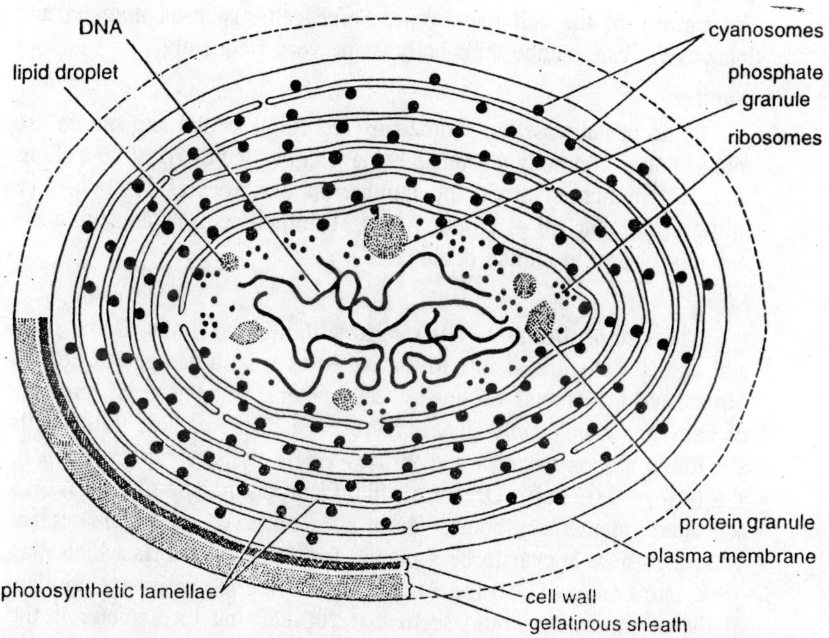

Fig. 1.4. A prokaryotic cell of cyanobacteria.

Eukaryotic Cells

Eukaryotic cells may be acellular organisms, such as protozoans and acellular algae, or they may be cells that make up the tissues and organs of multicellular organisms. Though the eukaryotic cells have different shape, size, and physiology but all the cells are typically composed of plasma membrane, cytoplasm and its organelles, viz., mitochondria, endoplasmic reticulum, ribosomes, Golgi complex, etc., and a true nucleus.

Morphology of Eukaryotic Cell

Shape

Cell shapes are almost as numerous as cell types: there is not typical shape. The cells of certain unicellular forms, such as *Amoeba, Diatoms, Acetabularia* and bacteria exhibit a number of shapes. But generally the cells are rounded or spherical. Besides this the cells like oval, cuboidal, cylindrical, flat, discoidal, polygonal have also been observed. The shape of cells depends mainly on functional adaptations and partly on the surface tension and viscosity of the protoplasm, the mechanical action exerted by the adjoining cells and the rigidity of the cell membrane. Some cells, such as *Amoeba* and leucocytes, can change their body shape very frequently.

Number

Some organisms like protozoans are single celled and others are multicellular. The body of human being is composed of about 26 trillions of cells. In human blood the number of erythrocytes is about five million per cubic ml of blood. About 10 billion neurons constitute the nervous system in human being.

Size

The size of different cells ranges within broad limits. Some plant and animal cells are visible to the naked eye, such as eggs of certain birds have a diameter of several centimeters. But the great majority of cells are visible only under microscope. The smallest living cells are found among bacteria and viruses where the size ranges 0.1 μ to 1 μ (1 μ = 0.01 mm). Organisms like Pleuropneumonia-like organisms so called "elementary bodies" have been observed with diameters of 100 mμ. These appear to be a resting from of the bacteria which may grow into bodies of 250 mμ in diameter during the active metabolism of the organism. It would seem that 200-250 mμ in diameter is the lower limit for the size of inactive, living cells. This lower limit may be set by the minimum number and size of the component necessary

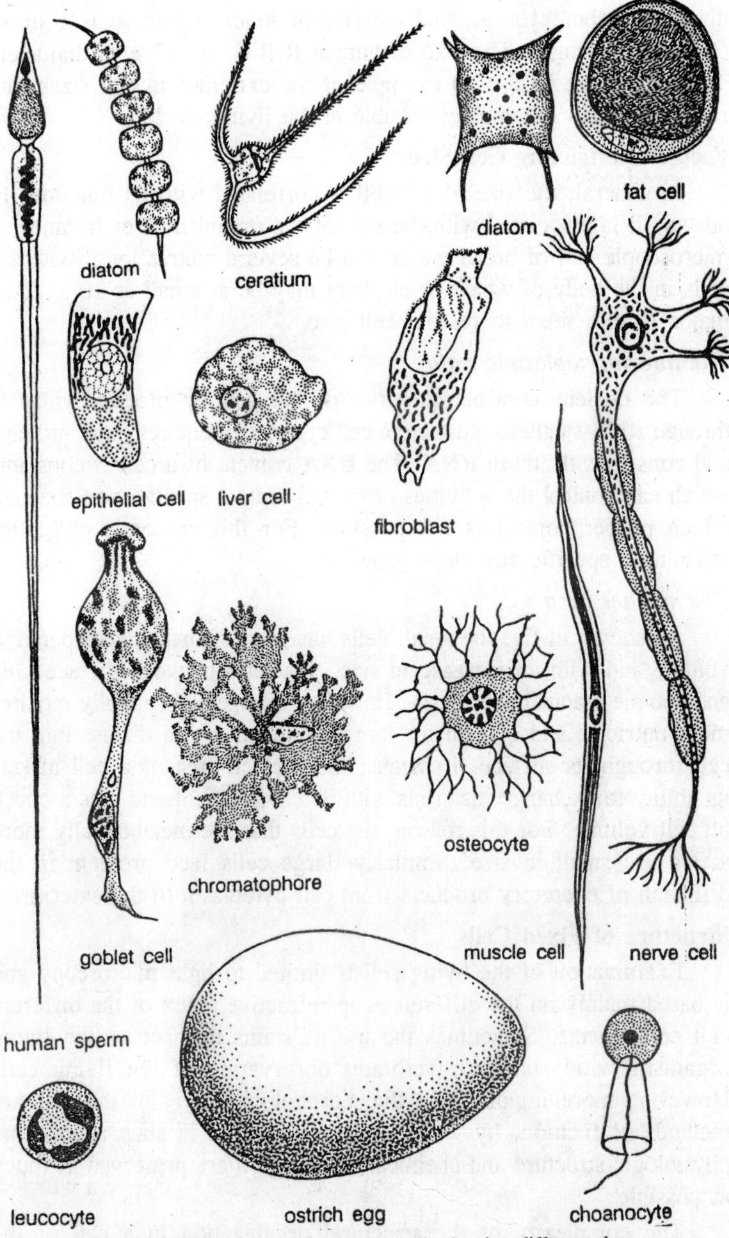

Fig. 1.5. Various types of eukaryotic cells showing different shapes.

for independent cellular existence. Usually the vast majority of cells lie in the range of .5 to 20 μ in diameter. The length of acellular diatoms is about 100 μ. And the size of *Amoeba proteus* is 1 m.m. (1000 μ) in length. The size of human R.B.C. is 7-8 μ in diameter. The long nerve is also an example of the extremes in cell sizes, the range of which is quite remarkable in the living world.

Factors Restricting Cell Size

In general, the size of the cell is correlated with its function. In no way it is associated with the size of the organism. For instance, a microscopic cell of bread mould can be several metres long, whereas cells in the body of whale or elephant may be as small as 10 μ. Two major factors seem to restrict cell size.

The nucelocytoplasmic ratio

The nucleus controls activities of the cell by protein synthesis through RNA synthesis. In a large cell cytoplasm requires more proteins and consequently more RNA. The DNA content of a cell is constant, which can control the activities of the cell upto a specific size, beyond which proper control is not possible. For this reason a cell after attaining a specific size stops growing.

The surface area

As shown in fig, the small cells have more surface area per unit volume and with an increase in size, this ratio between surface area and volume gradually decreases. However, a large cell naturally requires more nutrients and oxygen for its metabolism, which diffuse into the cell through its surface. It means, the surface area of a cell affects its ability to exchange materials with its environment and puts a check on cell volume. For this reason, the cells that are metabolically more active are small in size. Similarly, large cells face problem in the diffusion of excretory products from cell cytoplasm to the exterior.

Structure of Fixed Cells

Examination of the living cell is limited to light microscopy and is based mainly on the differences in refractive index of the different cell components. Sometimes the use of stains that act on the living organism (vital staining) facilitates observation of the living cell. However, more important in the morphologic study of the cell are methods of fixation, by which cell death results in such a way that physiologic structure and chemical composition are preserved as much as possible.

The complexity of the structural organization in a cell of the higher plants and animals is most impressive. Although there are great differences between the primitive forms of life, and the higher plant

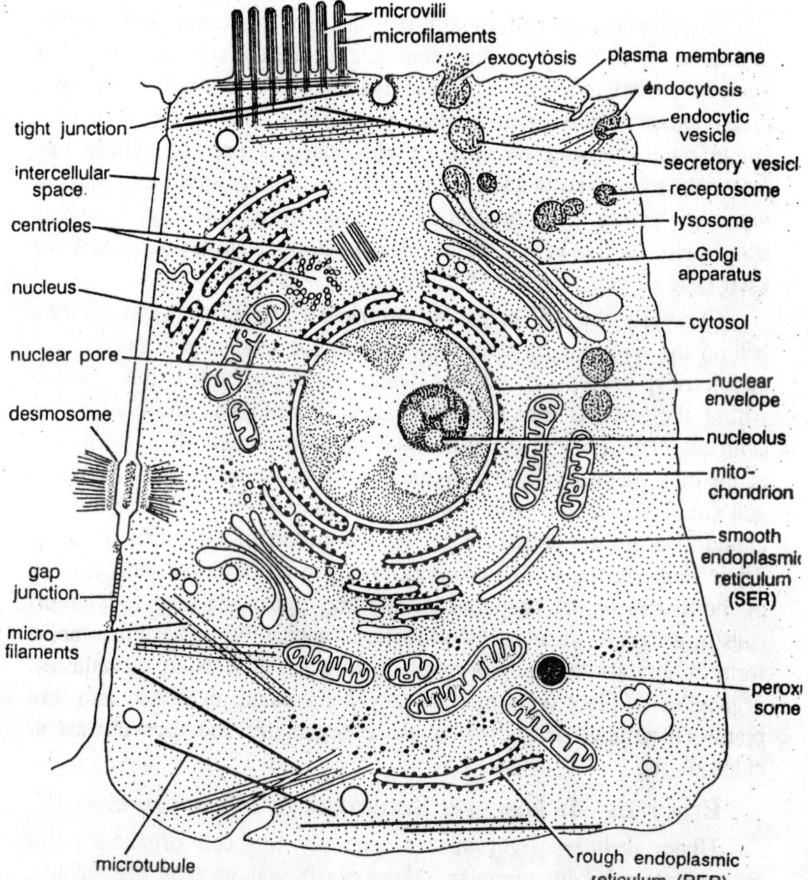

Fig. 1.6. Ultrastructure of a typical animal cell as seen in the electron microscope.

and animal cells, the similarities between primitive and advanced cells are also notable. Eukaryotic cells are characterized by a true nucleus with a nuclear membrane or envelope which divides the cell into two main compartments: nucleus and cytoplasm. The cytoplasm is turn is limited by the plasma membrane. In a plant cell the plasma membrane is covered and protected on the outside by a thicker cell wall through which there are tunnels, the ***plasmodesmata,*** by which the cell intercommunicates with neighbouring cells via fine cell process. In animal cells, parts of the plasma membrane are covered by a thin layer of material, which is generally described as the extraneous coat of the plasma membrane. The co-called *basement membranes* correspond to this extraneous coat.

In every stained cell usually one nucleus is present. It is central organelle of cell and is the main site of hereditary material. The nucleus is surrounded by a nuclear membrane but in bacteria and blue green algae it is without nuclear membrane. Within the nucleus a single nucleolus and network of chromatin is present. These two materials remain suspended in the nucleoplasm. A small cytoplasmic organoid 'centriole' is present in the majority of animal cells and in some cells of the lower plants. The position is generally fixed for each type of cell.

Generally the cell center or centriole occupies an axial position behind the nucleus. In some cells, it has the tendency to occupy the geometrical center i.e., in the leucocytes it occupies this position within the nucleus which is horseshoe shaped. The morphology of centriole changes with the functional state of the cell. It is cytoplasm of the cell, there are several organelles out of them, only mitochondria and Golgi bodies are visible under light microscope. Mitochondria or chondriosomes are scattered in the cytoplasm of all kinds of cells. Their shape in the fixed cells depends upon the fixation and handling of the tissues before preparation they may appear as long rods, short rods or as small granules. Golgi material appears in its very prominent form in nervous and secretory cells in animals. Plastids are cytoplasmic organelles of plant cells may or may not contain pigments. In green plants chlorophyll is present in these organelles and thus termed as chloroplasts.

Electron Microscopic Structure of a Typical Cell

Under ordinary light microscope only few cell organelles like mitochondria, Golgi complex, chloroplasts and nucleus are visible. However, under electron microscope, several other cytoplasmic organelles such as endoplasmic reticulum, ribosomes, lysosomes, nuclear membrane, plasma membrane, etc. are seen.

Some important cellular components are being described here under the following heads:

Plasma Membrane

The outermost boundary of the animal cell is called the 'plasma membrane'. It is invisible under a light microscope but under electron microscope, it appears to be composed of two dense layers which are called the outer dense layer and inner dense layer. Both these layers are about 20Å in thickness, and are separated from each other by a less dense area of about 35Å in thickness. By this way the total thickness of plasma membrane is about 75Å. Before the advance of electron microscope people used to think of the plasma membrane has been stretched tightly over the cell. But now we know that a single plasma membrane can have as many as 3000 microvilli. Another peculiarity of the structure of the membrane is that, it remains connected with the endoplasmic reticulum.

Plasma membrane is lipo-proteinaceus in composition.

Functions

(i) Gives an identity to the cell.

(ii) Protects the cell organelles.

(iii) Helps in digestion through phagocytosis and pinocytosis.

(iv) Being permeable to water, water can pass through it.

(v) Help in the transport of ions.

(vi) Helps in the formation of cytoplasmic organelles.

Mitochondria

The term mitochondria was, for the first time used by *Benda* (1902) to designate the filamentous and granules of cell cytoplasm. Most of the knowledge related to mitochondrial fine structure is due to the studies of *Palade* (1952, 55, 56) and *Sjostrand* (1955). Due to these studies, now mitochondria are defined as the double walled structures of cell organelles within which the membranous structures called *cristae* are present. The outer chamber of mitochondrion is made up of two dense layers and a less dense layer having the total thickness of 140 Å to 180Å while the outer and inner dense layers vary in their measurements from 35Å to 60Å in thickness depending upon the physiological state of the cell. The less dense layer is 40-47Å in thickness.

The matrix of the mitochondria may have some fine opaque granules of various size. *David E. Green* (1964) observed in a very highly magnified electron microscope that the inside of the inner membrane and the outside of outer membrane are covered with very small particles. They measure about 90-100Å in diameter. The particles on the outer membrane differ from those on the inner membrane. The former generally appear as simple spheres packed closely together which give to the surface of the mitochondrion, a rough or pimpled appearance. The particle on the inside of the inner membrane has the more complex configuration. In some cells, but not all, this particle is seen to have a base, a stalk and a spherical head. Mitochondria represent the power generating centres of the cell. They can bring about oxidation of the intermediates of carbohydrates, fat and amino acid metabolism in the presence of O_2 releasing energy. The released energy is trapped in the form of adenosine triphosphate (ATP).

Golgi Complex

Golgi comples have probably been first observed earlier in 1867 by *La. Vallete St. George,* but any how credit goes to *Camillo Golgi* (1898) who described it as internal reticulate apparatus' and after it, was named *Golgi apparatus.* It is variously called Golgi element, Golgi complex, Dalton's complex etc. Each Golgi complex is composed of many lamella, tubules, vesicles and vacuoles. The membranes of Golgi complex are of lipoproteins and these are supposed to be originated from the membranes of endoplasmic reticulum. The function of the Golgi complex is storage of proteins and enzymes which are synthesized by ribosomes and transported by endoplasmic reticulum to them. Further the Golgi complex has most important secretory function. It secretes many secretory granules and lysosomes. In plant cells the Golgi complex is known as dictyosome and secretes necessary materials for cell wall formation during cell division.

Lysosomes

The lysosomes (Gr. *lyso;* digestive+*some;* body) are also found scattered in the cytoplasm. By gently centrifuging lysosomes can be separated from other cell organelles. They are usually spherical but may be irregular in shape, covered with a membrane of lipoprotein. Inside structure is always variable. Lysosomes contain hydrolysing enzymes, such as acid phosphatase, acid ribonuclease, acid deoxyribonuclease and cathespins. But they do not have oxidative enzymes. This property distinguishes them from the mitochondria. Lysosomes are polymorphic in nature and are commonly known as suicide bags and are responsible for postmortem changes in cells.

Endoplasmic Reticulum

The term "Microsome" was introduced to the modern cytology by *Claude* (1943) to represent one of the submicroscopic cellular components isolated by centrifugation. These are now known to be the broken parts of the endoplasmic reticulum after the studied of *Kollman* (1953). *Porter* (1945) was the 1st man to describe their electron microscopic structure in cultured cells. He described them as lace-like reticulum. These are membrane bounded sacs in the form of double membrane or cisternae. The term cisternae was introduced by *Sjostrand* (1953) to describe long or elongated rod like parts of the endoplasmic reticulum measuring 50-200Å. *Wiess* (1953), described the vesicles of the endoplasmic reticulum which usually has a diameter of 25-500Å while *Bradfield* (1913) came across of another type of endoplasmic reticulum termed them as tubules, they measures from 50 to 100Å in diameter.

On the basis of presence or absence of ribosomes or RNP particles the endoplasmic reticulum (E.R) is distinguished in two varieties. The *rough walled endoplasmic reticulum,* and *smooth walled endoplasmic reticulum.*

Rough Walled Endoplasmic Reticulum

Rough walled endoplasmic reticulum (R.E.R.) is that variety which bears the ribosomes at the external surface of the cisternae. The distribution of the ribosomes can be circular, spiral, or rosette type. These particles may be induced to leave the surface of the cisternae without damaging the structure of the cisternae or the capacity to associate again with the particles. Usually the rough walled endoplasmic reticulum is richly distributed in those cells which are engaged in the synthesis of the protein.

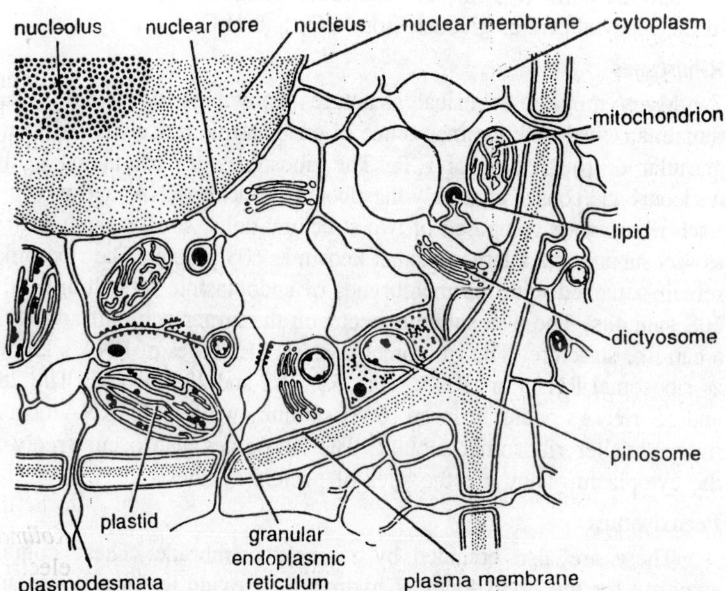

Fig. 1.7. Ultrastructure of plant cell (showing small portion of cell).

Smooth Walled Endoplasmic Reticulum

The other large division of the system owes its identity in parts of the absence of the particles and therefore commonly referred to as the smooth surfaced a granular form. Like the R.E.R., the S.E.R. shows a characteristic morphology which is tubular rather than cisternae. The tubules have the diameter nearly of 50-100 mm. As studied by *Fawcett* (1960) S.E.R. is richly distributed in those cells which are engaged in the synthesis of steroids.

Continuity between the S.E.R. of a particular cell type possess a characteristic pattern of structure. Thus it is typical for rat liver cells to show groups of eight or ten slender profiles in parallel array.

Plastids

Plastids are present only in plant cells (not in animal cells). These are double-membrane structures relatively larger than the mitochondria. Some of the plastids are colourless and store starch (Leucoplasts). Other contain various pigments and impart different colours (red, brown, yellow, etc.) to different parts of the plant. These are collectively called chromoplasts. Those containing green pigment—cholorophyll, are called choloroplasts. These contain several enzymes in their matrix necessary for photosynthesis. The chloroplasts capture light energy of the sun and convert it into the chemical energy—ATP, which is used in the biosynthesis of glucose from CO_2 and H_2O.

Ribosomes

Many minute, spherical structures known as *ribosomes*. They remain attached with the membrane of endoplasmic reticulum and forms granular or rough type of E.R. The ribosomes are originated in the nucleolus and consist of mainly the ribonucleic acids (RNA) and proteins. Each ribosome is composed of two structural units, small subunit known as 40S subunit and a larger subunit known as 60S subunit. The ribosomes remain attached with the membranes of endoplasmic reticulum by the 60S subunits. The 40S subunits occur on the larger sub unit and form a cap-like structure. The ribosomes consists of 3 types of RNA's known as ribosomal RNAs or rRNA, viz., 5S, 18S and 28S rRNAs. The 28S and 5S rRNAs occur in large (60S) subunit, while 18S rRNA occurs in the smaller ribosomal subunit. The ribosomes also occur freely in the cytoplasm. They are the sites of protein synthesis.

Peroxisomes

These are also bounded by a single membrane. These contain enzymes for the breakdown of hydrogen peroxide to form water and oxygen. These are protective organelles of the cell.

Cilia and Flagella

The cells of many unicellular organisms and ciliated epithelium of multicellular organisms consists of some hair-like cytoplasmic projections outside the surface of the cell. These are known as cilia or flagella. They help in locomotion of the cell. The cilia and flagella consists of nine outer fibrils around the two large central fibrils. Each outer fibrils consists of two microtubules. The two microtubules of each peripheral pair or *doublet* remain morphologically distinct. The central fibrils are separate but remain enveloped in a common *central sheath.* Cilia and flagella are envolved in locomotion, feeding, cleansing etc.

Vacuoles

These structures are fluid-filled "bubbles" found in the cytoplasm. Vacuoles are bounded by a membrane, called *tonoplast,* which is structurally similar to, if not identical with, a cell-membrane. Inside the vacuoles, food materials or wastes are found.

Vacuoles are consipicuous in plant-cells but they are found less in animal cells. Vacuoles are absent in actively dividing cells. In a young plant-cell, there are many vacuoles but, when a cell matures, they unite to form a single, large, central vacuole.

Guilliermond (1941) was of the opinion that vacuoles are colloidal with a strong imbibing capacity. Vacuoles arise *de novo* i.e., spontaneously formed from pre-existing vacuoles. In many protozoan, vacuoles are formed at the bottom of the gullet. In these vacuoles, food-materials, such as bacteria, are collected. Some vacuoles are formed by cell-drinking. This type of vacuoles are called *pinosomes.* Food-vacuoles formed by evagination of cell-membrane surrounding food-particles with some fluid, are called *phagosomes.* In animals cells, the vacuoles are formed by a lipoprotein membrane that helps in storage and transport of materials and in the maintenance of internal pressure of the cell. Vacuoles are not a living part of the protoplasm but are a storing place of reserve materials and waste products.

Crystals and Oil Droplets

In certain plant cells the food or waste material is found deposited in the form of crystals. The oil droplets are round in both plant and animal cells, which appear as tiny glistening white spheres. These serve as a reserve supply of energy rich fuel.

Nucleus

It is the most vital part of cell which is invariably present in all the cells of higher animals and plants. Normally one nucleus is present in a cell but more than one may be present in certain cells. Structurally it is enveloped by a porous *nuclear membrane* that separates it from the surrounding cytoplasm. It has the characteristic permeability. During cell division, it breaks up and soon after the mitosis it is reconstituted. The inside of nucleus is filled with *nuclear sap* or *karyolymph.* In nuclear sap remain embedded other nuclear constituents. The *chromatin* is the Feulgen positive material in the interphase nucleus which during division becomes the definite chromosomes. In short it represents the chromosomes substance.

The sex-chromatin are the specific heterochromatic bodies, found generally near the periphery of the nucleus. In nucleus usually one sex chromatin body is found but more than one can also be observed. *Nucleolus* is relatively large, generally spherical ball-like body. It has a high concentration of RNA and thus is an important structure involved in protein synthesis. Its number in a nucleus vary in different cells and depend on either the species or the sets of chromosomes. The nucleus co-ordinates many cell activities. DNA molecules are the source of coded instructions, which pass to RNA molecule which then act as messenger, moving out of the nucleus to one of the various structures of the cytoplasm. It also controls the cell reproduction. Here too, DNA is the source of coded instructions.

The Cytosol, Hyaloplasm or Ground Substance

The material that remains suspended under conditions sufficient to sediment even the ribosomes is referred to by biochemists as the *soluble fraction* or *cytosol.* This fraction usually contains some material lost from the organelles in the course of homogenization and centrifugation. However, it also contains other molecules, including some enzymes, that are not known to be part of the intracellular structures indentifiable in the microscope. Such molecules are usually thought to derive from the *hyaloplasm* (also called ground substance, cell matrix, or cell sap), which was originally defined by microscopists as the apparently structureless medium that occupies the space between the visible organelles. As will be evident in subsequent chapters, with the improved techniques less and less of the cell appears structureless.

Most investigators continue to believe that there does exist in the cytoplasm a "soluble phase"—a solution containing inorganic ions and small molecules, materials in transit from one cell region to another and dissolved or suspended enzymes that function as individual molecules or perhaps or parts of small groups of molecules rather than as components of large, microscopically recognizable organelles. But there is a lively debate going on over proposals that many components once though to be suspended in this solution are actually bound, perhaps loosely, in structured arrangements that do not survive the disruption involved in cell fractionation.

2

CHEMICAL ENVIRONMENT OF CELL

Protoplasm is a highly complex mixture of some elements and compounds found in the bodies of living beings. The protoplasm is variously known as the *living matter*, *living substance* or *physical basis of life*. It is the basic fundamental substance exhibiting all the vital processes of the cell. The protoplasm was first observed by *Corti* in 1772. In 1835 *Dujardin* a Frenchman described it as a soft and gumy substance and named it as *scarcode* i.e. *flesh*. *Purkinje*, a Bohemian physiologist (1839), was the first biologist, who gave the name *protoplasm* to this living substance. *Hugo von Mohl,* a German botanist, in 1839, also suggested the name *protoplasm* for the granular and viscous substance found in plants similar to that found in the animals. The above-mentioned German botanist popularized the word protoplasm as a name given to the living matter found in plants and animals. Protoplasm is the most complex and interesting substance. It is not to be thought of a chemical compound but rather as very complex organized system. Protoplasm varies somewhat in its nature from cell to cell and from organism to organism, but basically it must be the same, as evidenced by its common manifestations of metabolism, growth, reproduction and by some other peculiarities.

PHYSICAL NATURE OF PROTOPLASM

Different workers have proposed different theories to explain the physical nature of protoplasm as given under the following heads:

Granular Theory

This theory was propounded by *Altmann* in 1893. According to this theory, protoplasm consists of numerous tiny granules as shown in *Amoeba. Henle, Maggi,* etc., considered these proto-plasmic granules

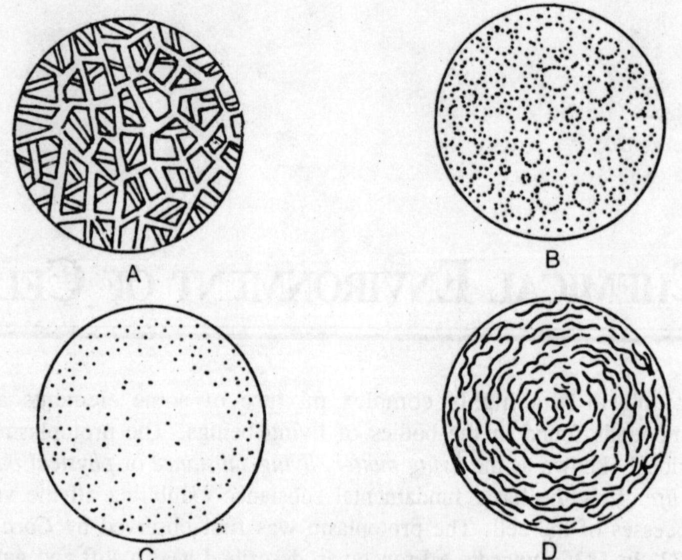

Fig. 2.1. Physical appearance of protoplasm. A—Reticular, B—Alveolar, C—Granular, D—Fibrillar.

plastidules. *Altmann* reconginzed them as "elementary organisms," or *bioplasts* (or cytoplasts).

Alveolar Theory

The alveolar nature of protoplasm was suggested by *Butschilli* in 1892. According to him, protoplasm consists of many suspended droplets or alveoli or minute bubbles, resembling to foam of emulsion.

Fibrillar Theory

This theory was put forward by *Flemming*. According to him, protoplasm consists of fibres embedded in the inner mass of matrix. The fibrillae are called mitome or spongioplasm and ground substance is termed paramitome or hyaloplasm.

Reticulate Theory

This theory was proposed by *Klein, Cornoy* etc., suggesting that the protoplasm consists of a network or reticulum of fibres in the ground substance.

Colloidal Theory

Proposed by *Wilson* in 1925. According to this theory the protoplasm having any of the aforesaid appearances is always a fluid-colloidal system having various chemical inclusions in *gel phase*. A gel is a semi-solid condition which presents jelly-like appearance. Here the molecules are held together by various bonds depends upon the imature and strength. By absorbing water the gel changes to more liquid-like phase. This is known as *sol* and the process is termed as *solation.* The sol can stream easily and by losing water again changes into gel-state. Both these states of colloidal matrix are interchangeable according to the various physiological, mechanical and biochemical activities of the cell. This represents phase reversal in the colloidal system. These sol-gel conditions of colloidal system are the basis for the mechanical behaviour of protolplasm.

Properties of Protoplasm

Cohesiveness

The various particles or molecules of protoplasm are adhered with each other by forces, such as Van der Waal's bonds, that hold long chains of molecules together. These Van der Waal's bonds are weak and non-specific forces between non-polar groups of atoms. This property varies with the strength of these forces.

Contractility

This property is significant in various stomatal operations in plants. The contractility of protoplasm is important for the absorption and removal of water as they generally occur in protoplasm.

Electrical Charge

Protein molecules repel each other because their overall charge is similar. All molecules are either positively charged or negatively charged. However, if the molecules approach one another close enough so that valency forces can act, then they may be attracted to each other.

Precipitation

Addition of certain amount of electrolyte in a colloidal suspension causes its dispersed particles to colloide, aggregate and finally to precipitate as suspension. For example, the addition of HCl to a colloidal system of arsenic sulphide causes precipitation.

Viscosity

The viscosity of the ground substance of the cell varies greatly. It may be as low as that of water, or may be very high in the gelating cytoplasm of pseudopodia of *Amoeba.*

Streaming Movement or Cyclosis

The protoplasm exhibits various sorts of streaming movements inside the cell. These have been studied in *Amoeba, Paramecium* etc. The movement involves only the localized portions with no visible changes in the protoplasm. No complete explanation of this has been given as yet. But it is seen that its rate depends upon the rate of cell metabolism. It is due to the fact that the energy is supplied by respiration.

Amoeboid Movement

The amoeboid movement as exhibited by *Amoeba* and other protozoans involves the movement of entire protoplasm of the cell, where the cytoplasm moves as one mass carrying the various inclusion with it. It is due to the continuous change of gel to sol and sol to gel.

Brownian Movement

It is characterized by the zigzag motion of suspended colloidal particles, occurring due to the bombardment of one particle or molecule by other. This type of movement of particles was first of all observed by *Robert Brown* in 1827 in the colloidal solution and hence such movements are known as Brownian movements. The higher the temperature, more rapid the movement and thus viscosity of cell is decreased. This means that high viscosity indicates a more gel-like state of protoplasm and low viscosity, a more sol-like condition.

Tyndall Effect

Colloidal particles of protoplasm have the property of scattering light. When a beam of light is passed through a colloidal solution it becomes visible. This is the Tyndall effect. A colloidal solution of proteins in water shows a typical Tyndall cone.

Adsorption

The tendency of particles, molecules or ions to adhere to the surface of certain solids or liquids is known as adsorption and is exhibited by the particles of colloidal system. The phenomenon helps the matrix to form protein boundaries.

Biological Properties

Protoplasm has all the biological properties of a living organism. It is capable of nutrition, respiration, excertion, metabolism, growth and reproduction. It has the property of irritability, e.g., it responds to stimuli like heat, light and chemicals. It also has the property of conductivity, i.e., of conducting impulses produced by stimuli.

Chemical Nature of Protoplasm

For the detailed study the chemical components of the cell can be classified as inorganic (mineral salts) and organic (proteins, carbohydrates, nucleic acids, lipids and so forth) substance. Although the most prominent constitutent of protoplasm is water—the substance which gives protoplasm its characteristic structure is protein. Lipids are important in all membranes and carbohydrates serve as nutrient stores. The protoplasm of a plant or animal cell contains 75 to 85% water, 10-20% protein, 2-3% lipids, 1% carbohydrates and 1% inorganic material. The following Table 2.1. gives approximate active protoplasm.

Table 2.1. Chemical Analysis of Protoplasm

Substance	*Percent*	*Average molecules weight*	*Number of molecues in relation to protein*
Water	85	18	180
Protein	10	36000	1
DNA	0.4	10	—
RNA	0.7	4.0×10^4	—
Lipid	2	700	10
Other Organic matter	0.4	250	20
Inorganic substances	1.5	55	100

Water

Water is a substance that is justifiably called the *fluid of life*. It comprises over 90% of the chemical content of most organisms. In human body about 55% of the water (20-22 litres) is intracellular water and remains confined to the cells and the rest is found in extracellular fluids such as blood, tissue fluid and lymph. Water participates directly or indirectly in all metabolic reactions. The biologically active conformations of macromolecules and arrangement of phospholipids in the lipid bilayer of membranes are dependent on water. Water helps to keep minerals ionised in body fluids. It ionises itself to provide hydrogen ion (H^+) concentration to body fluids. It also helps in maintaining the constancy of the internal environment of an organism.

Water is a remarkable compound with unique properties that result from its molecular configuration and hydrogen bonding. Some important properties of water are as follows.

Polarity

The water molecule is composed of two hydrogen atoms covalently bonded to one side of an oxygen atom. Since the mean angle (105°) between the hydrogen atoms is not rigid, water can absorb large quantities of heat and be subject to other physical stresses without breakage of the bonds. Water is a polar molecule and as with other polar molecules it has a surface charge. Water is a dipolar substance in that the hydrogen pole is positively charged and the other pole is negatively charged due to electrophilic (electron attracting) properties of oxygen. The polar nature of the water molecule causes salts to be held in solution through charge interaction (ionization; salt dissolved in water exist in the form of positive and negative charges).

Cohesion and adhesion

Because of the asymmetrical distribution water molecules associate with each other (cohesion) and 'wet' other substances (adhesion). The attraction of the positive hydrogen atom of one water molecule for the negative oxygen atom of another water molecule results in a *hydrogen bond*. Hydrogen bonding of water produces a dipolar molecule and favours the formation of a lattice-like structure that enables the packing of many atoms into a small area and stabilizes the molecular structure of water. Fluidity of water is maintained by very rapid formation and dissociation of hydrogen bonds between water molecules.

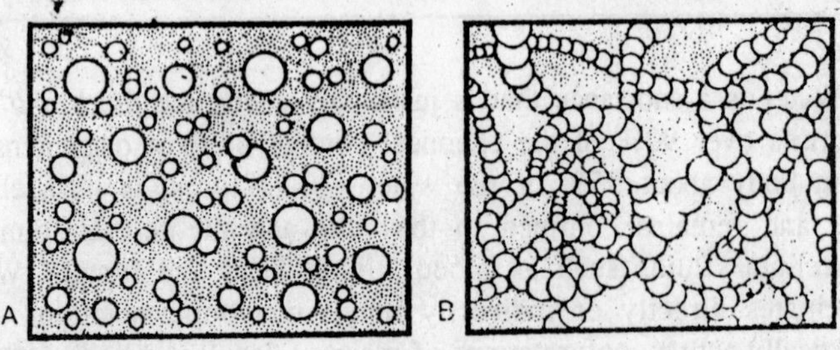

Fig. 2.2. Two water molecules showing polarity and formation of a hydrogen bond between them.

Latent heat and specific heat

The presence of hydrogen bonds is directly responsible for the high heat of fusion, high specific heat, and high heat of vaporization of water. The energy required to break hydrogen bonds for melting of ice, heating of water, and evaporation of water is considerably greater than the energy needed to overcome the Van der Waal's forces that are normally found in the weak association of molecules of ethane, ether and benzene. Due to high specific heat, water affords protection against sudden temperature changes in living organisms. High latent heat of evaporation of water causes elimination of excess heat through evaporation of sweat. This maintains a constant body temperature.

Density of freezing properties

The density of water decreases below 4°C and ice therefore tends to float. It is the only substance whose solid form is less dense than its liquid form. The fact that water below 4°C tends to rise helps to maintain circulation in large bodies of water. This may result in nutrient cycling and colonisation of water to greater depths.

Solvent action

A property of water that is very important to the living cell is its solvent action. Because it forms a solution with a vast array of compounds, water is sometimes referred to as the *universal solvent.* The solvent action of water is an effect of its ability to form hydrogen bonds due to the asymmetrical distribution of its charges. In solution with water, compounds like sugars, alcohols, and amino acids—which contain oxygen atoms, hydroxyl (-OH) groups, amino ($-NH_2$) groups—form hydrogen bonds with the molecules of water.

The solvent action of water is of tremendous importance for the living system. The essential elements necessary for normal growth, the compounds necessary for energy transfer and storage, and the components of structural compounds all require water as a translocation and reaction medium. Indeed, physiological processes in dilute solutions and suspensions; and the reactions are, therefore, under control of the physical and chemical laws that govern the activities of dilute solutions and suspensions.

Inorganic Compounds

A number of inorganic salts occur both in free as well as in the ionised state. The elements that occur in quantity in protoplasm are oxygen, carbon, hydrogen and nitrogen. Always present but in much smaller amounts are potassium, phosphorus, calcium, sulphur, magnesium and iron. These ten are generally referred to as *essential elements*. Chlorine and sodium are necessary components of most animal protoplasm but are apparently not essential for plant protoplasm. Copper, boron, iodine, maganese, zinc and several other elements which are present in all protoplasm but only in minute quantities, are called *trace elements*. This term must be used with care, because these elements are still important even though they are present only in very small quantities. The essential elements found in human protoplasm are listed in table 2.2.

The elements are almost always present in protoplasm in the form of chemical compounds rather than elements. Many of these compounds inorganic salts, which are usually in solution. These salts have numerous functions. They serve as source of elements to be built into other compounds, and some act as *buffer* in maintaining the proper acidity or alkalinity in the protoplasm. Because salts in solutions are electrical conductors, they also function in some way in connection with the electrical properties of protoplasm, however, this is not well understood at present.

Table 2.2. Showing Percentage by Weight to the Element in Protoplasm

Elements	*Symbol*	*Percent*
Oxygen	O	65.04
Carbon	C	18.24
Hydrogen	H	10.05
Nitrogen	N	3.15
Potassium	K	1.60
Phosphorus	P	0.84
Calcium	Ca	0.25
Sulphur	S	0.20
Magnesium	Mg	0.04
Iron	Fe	0.01
Chlorine	Cl	0.27
Sodium	Na	0.26

ORGANIC COMPOUNDS

Proteins

Protein is an indispensable constituent of the diet because it is the only source of the amino acids. Amino acids are needed to built up new tissue during he period of growth; to maintain the structure of every tissue cell including its content of protein—containing enzyme systems; to provide raw material for the manufacturing of digestive enzymes and certain hormones; and to maintain the normal concentrations of plasma proteins and haemoglobin.

The proteins are macromolecules of very high molecular weight. The polymer molecules of protein are formed by linking together of a number of amino acids. About 20 odd different kinds of amino acids are known. Each amino acid has an amino (NH_2) and a carboxyl (COOH) group. A bond is formed between the amino group of one amino acid with the carboxyl group of another amino acid. The bond is called peptide bond. With the help of such peptide bonds a variety of amino acids can form a long chain molecule called polypeptide.

The nutritional value of a protein depends upon its amino acid composition. Synthesis of amino acids from organic keto-acids is of common occurrence in bacteria and plants. Animals can synthesize only about half of the naturally occurring amino acids (20). Thus there are two categories of amino acids:

Essential amino acids

The amino acids which cannot synthesized by an organisms i.e. they must be obtained from the dietary proteins are called essential amino acids. These are *arginine*, *histidine*, *isoleucine*, *leucine*, *lysine*, *methconine*, *theronine*, *phenylalanine*, *tryptophan* and *valine*.

Non-essential amino acids

Although these amino acids are required by the animal as they are found in the protein of the tissues, but they can apparently be synthesized from alpha-keto acids, by the process of amination. These amino acids are *tyrosine*, *serine*, *alanine*, *asparagine*, *proline*, *hydroxyproline*, *asparatic acid*, *glycine*, *glautamine*, *glutamic acid* and *cystine*.

People who eat diets lacking only one of the essential amino acids are unable to synthesize normal body proteins and become protein deficient, despite having a normal nitrogen intake.

Most animal proteins, are *complete or first class proteins;* that is they contain all the necessary amino acids in appropriate proportions for ultilization by man as eggs, meat, kidney, fish, liver, poultry, milk etc.

Plant proteins are *incomplete* or *second class proteins,* since they lack certain essential amino acids. as. cereals, nuts, legumes etc. Thus people living on a strict vegetarian diet may become protein deficient.

Associated with protein deficiency is *Kwashiorker,* which affects millions of children in developing countries. It must commonly occurs in infants after weaning and particularly when given an inadequate and predominantly carbohydrate diet. Many of the children also have an inadequate energy intake and the disorder is known as *protein-energy malnutrition* (PEM). The affected child stops growing and loses weight. The skin and hair may become depigmented, and oedema associated with low plasma levels of albumin often develop. Untreated Kwashiorker is often fatal.

STRUCTURE OF PROTEIN

The critical determination of biological function of a protein is its conformation, which is defined as the three-dimensional arrangement of the atoms of a molecule. Four basic structural levels are assigned to proteins. These are:

Primary Structure

Arrangement of amino acid in a polypeptids chain is the primary structure of a protein. Amino acid sequence of the smallest protein known, that is of insulin, was studied by *F. Sanger* (1954). Insulin consists of two chains, A and B, joined together by two disulplide bonds (A chain contains 21 and B chain 30 amino acids). Normal human haemoglobin consists of four polypeptide chains (2α and 2β) to be held together by non-covalent forces. Each α-chain is composed of 141 amino acids, and each β chains 146 amino acids.

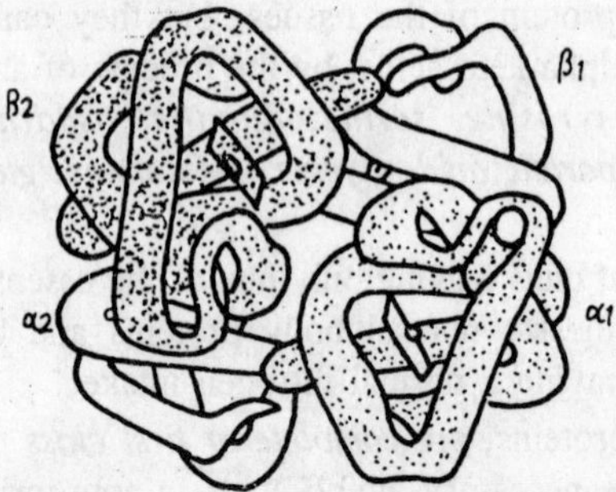

Fig. 2.3. Structure of haemoglobin; the molecule consists of four chains: two α-chains and two β-chains.

Secondary Structure

The polypeptides are held together to give rise a definite shape (helical pattern) of the protein molecules by hydrogen bonds. Hydrogen bonds can be broken by many physical and chemical treatments, for example, by excessive heat, pressure, *pH*, electricity, heavy metals and other agents that changes the environment of protein.

The most common type of secondary structure is α-helix. This is formed by the bending of polypeptide chain to form hydrogen bonds in the same chain. The bending of the chain is very regular and chain assumes the shape of a helix.

Pauling and *Corey* have considered an alternate type of secondary structure. In this case the polypeptide chains are lead linearly either parallel or antiparallel with respect to one another. This is called β-structure of β-helix.

Tertiary Structure

The arrangement and inter-relationship of the twisted chains of protein into specific loops and bends in called the tertiary structure. Such a structure enables the proteins to form specific layers, crystals or fibres. The tertiary structures is maintained by hydrogen bonds disulphide bonds, Van der Waals interaction, hydrophobic interactions and ionic bonds. The tertiary structure is important and found in globular proteins. If this structure is disrupted, the biological activity of protein would be lost.

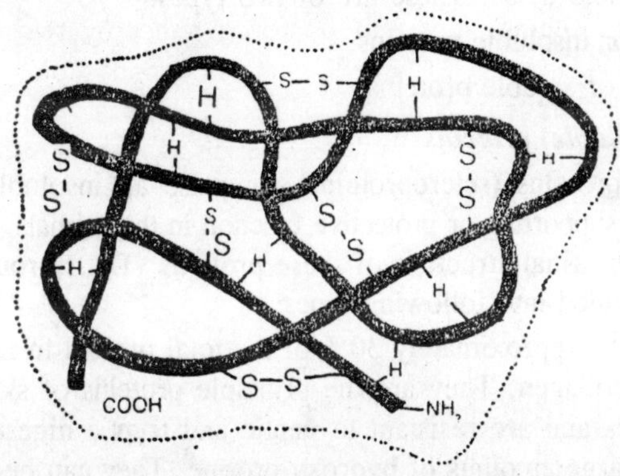

Fig. 2.4. Tertiary structure of a hypothetical protein molecule; note the extensive folding of the secondary structure which imparts a globular shape to the protein.

Quaternary Structure

This defines the degree of polymerisation of a protein unit. This structure is found in only oligomeric proteins. They are composed of

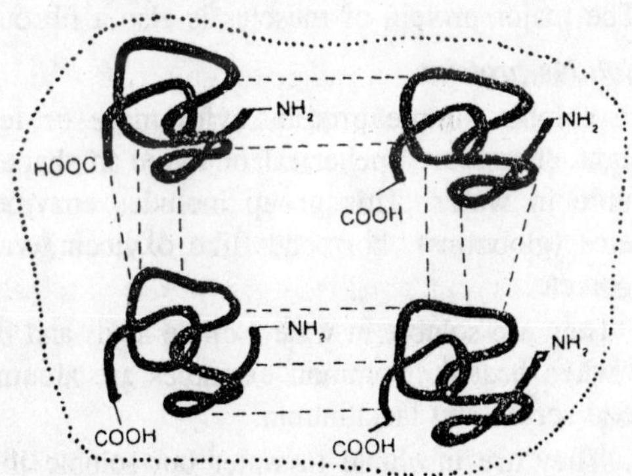

Fig. 2.5. Quaternary level of protein structure; note a diagrammatic representation showing four sub-units which are polymerized in the functional protein molecule.

subunit peptide chains linked together by any or all of the forces that can act between amino acid side chains. Haemoglobin is a fine example of quaternary structure and was studied by *Kendrew* and *Perutz* (1963).

Classification of Protein

The proteins are mainly classified into two—simple and conjugated proteins.

Simple Proteins

These are consisting mainly of peptide chains. On hydrolyses they yield only amino acids. These are of two types:

(A) Fibrous or insoluble proteins.

(B) Globular or soluble proteins.

Fibrous (insoluble) proteins

Fibrous proteins (scleroproteins) comprise all insoluble proteins which have a supporting or protective function in the animals. Secondary structure is the final structure of these proteins. The fibrous proteins can be subdivided into following types:

1. *Collagens*. Approximately 30% of the total protein in mammalian body is collagen. They are the principle proteins of skin, tendon and bones and are resistant to peptic and tryptic digestion. They contain large amounts of hydroxy-proline. They can be converted into soluble gelatin by boiling with water, dilute acids or alkalies.
2. *Elastins*. They are present in ligament, arteries and other elastic tissue. They cannot be converted into gelatin.
3. *Keratins*. They are present in animal skin, nails, hairs, horns, hooves, claws, feathers etc. they are derived from ectoderm.
4. *Myosin*. The major protein of muscles is also a fibrous protein.

Globular or Soluble proteins

These are soluble simple proteins with more or less definite molecular weight. These are speherical or ovoid in shape. They are generally soluble in water. This group includes enzymes, oxygen carrying proteins (globulins), hormones like oxytocin, vassopressin, insulin, glucogon etc.

1. *Albumins*. They are soluble in water, dilute acids and bases. They coagulate when heated. Common examples are albumins of egg white, blood serum and lactalbumin.
2. *Globulins*. They are insoluble in water but soluble in dilute salt solution. They are also coagulated when heated. Examples are serum globulin, fibrinogen, ovoglobulin in egg-yolk etc.

3. *Glutelins*. These are insoluble in water but soluble in dilute acids or base solution. They are commonly found in various plant seeds. Examples are wheat, cereals etc.
4. *Protamins*. These are basic proteins. They are very soluble, small, stable proteins and cannot be coagulated by heating. They are exceptionally rich in arginine amino acid. They are found in the ripe sperm cells of certain fishes.
5. *Histones*. These are water-soluble basic proteins. They coagulate on heating. These are found associated with nucleic acids in nucleoproteins. They are obtainable from the thymus, spleen and nucleated erythrocytes of birds.

Conjugated Proteins

The conjugated proteins differ from the simple proteins in that they consist of proteins combined with some non-protein substances. The term *prosthetic group* is generally used to designate such substances (non-proteinous). They are classified according to the nature of prosthetic group as follows:

1. *Nucleoprotins*. These are formed by the combination of nucleic acids with protein.
2. *Mucoproteins*. They contain carbohydrate bound to protein molecules. The protein components of *mucoids* are combined with large amount (more than 4%) of carbohydrate. Examples are ovomucoid-β from egg white; or osomucoid from blood serum.
3. *Glycoproteins*. The carbohydrate percentage is less than 4% in glycoproteins. The carbonhdrate in these is usually the mucopolysaccharids. Mucin of saliva, chorionic gonadotropins and some other hormones of pituitary such as follicle stimulating hormone and luteinizing hormones are glycoproteins.
4. *Lipoproteins*. They are compounds of lipid and protein. The protein part is water soluble and lipid unsoluble. Examples are cholesterol, phosphatidylglycerides, lipovitelline of egg-yolk and serum lipoporteins.
5. *Chromoproteins*. These are the proteins which have a matelloprophyrin or some similar substance. Examples are haemoglobin which contains a basic protein globin with an iron porphyrin as Heme. Haemocyanin, haemoerythrin, erythrocuorins are other examples.
6. *Phosphoproteins*. In these the protein molecule is linked to phosphoric acid when treated with dilute sodium hydroxide yield inorganic

phosphate. Examples are casein of milk, vitellin of egg-yolk and pepsin.

7. *Metalloproteins*. These proteins contain metals as an inherent part of their molecules. They include tyrosinase (Cu), carbonic anhydrase (Zn), arginase (Mn or Mg) etc.
8. *Flavoproteins*. The prosthetic group of flavin component remains permanently attached to the protein. Flavoproteins act as coenzymes which catalyze oxidation-reduction reactions.

Derived Proteins

These include the products obtained from proteins by the action of heat and other physical agents or by hydrolysis.

Denatured and Coagulated proteins

Upon treatment with certain chemical agents or heat treatment the proteins become insoluble on account of some sort of intramolecular rearrangement and it is said to have denatured and when it gets separated in the form of precipitate, it is said to have undergone coagulation.

Peptides

These include the various fragments of the proteins which are broken off during hydrolysis of the big protein molecules either the presence of the acids or enzymes. The larger fragments had previously been named as the proteoses and the smaller fragments as peptones. This terminology is being abandoned now.

Properties of Proteins

1. ***Colour and Taste.*** Proteins are colourless and usually tasteless. These are homogenous and crystalline.
2. ***Molecular weight.*** The proteins generally have large molecular weights ranging between 5×10^3 and 1×10^6. It might be noted that the values of molecular weights of many proteins lie close to or multiples of 35,000 and 70,000.
3. ***Colloidal nature.*** Because of their giant size the proteins exhibit many colloial propertis such as (a) their diffusion rates are extremely slow, (b) they may produce considerable light— scattering in solution, thus resulting in visible turbidity (Tyndall effect).
4. ***Denaturation.*** In refers to the changes in the properties of a protein. The process of denaturation is followed by coagulation. Denaturation may be brought by a variety of agents. The physical agents, include mechanical action, heat treatment, cooling and freezing operations, ultraviolet rays etc. The chemical agents include X-rays, acetone, alcohol (solvents), salicyclates etc.

5. ***Amphoteric nature.*** The proteins are amphoteric i.e., they act as acids and alkalies both. These migrates in an electric field and the direction of migration depends upon the net charge possessed by the molecule. The net charge is influenced by the *pH* value. Each portion has a fixed value of isoelectric point (pl) at which it will move in an electric field.
6. ***Solubility.*** It is influenced by *pH* solubility is lowest at isoelectric point and increases with increasing acidity or alkalinity.
7. ***Hydrolysis.*** The proteins are hydrolyzed by a variety of hydrolytic agents.

(a) *By acidic agents.* Proteins, upon hydrolysis which come HCl (6-12 N) at 100-110°C for 6-20 hrs yields amino acids in the form of their hydrochlorides.

(b) *By alkaline agents.* Proteins may be hydrolyzed with 2N NaOH. This process is less used as it is highly disadvantageous.

(c) *By proteolytic enzymes.* Under relatively mild conditions of temperature and acidity, certain proteolytc enzymes such as pepsin and trypsin hydrolyze the proteins. Enzyme hydrolysis is used for the tryptophan.

8. ***Reactions involving—COOH group***

(a) *Reaction with alkalies* (salt formation). The carboxylic group of amino acids release a H^+ ion and forms carboxylae (COO^-) ions. These may be neutralized by cations like Na^+ and Ca^{++} to form salts.

(b) *Reaction with alcohols* (*esterification*). With alcohols, corresponding esters are produced.

$$R\text{—}\underset{NH_2}{\overset{H}{\underset{|}{\overset{|}{C}}}}\text{—}COOH + HOC_2H_5 \xrightarrow[\text{catalyst}]{\text{Acid}} R\text{—}\underset{NH_2}{\overset{H}{\underset{|}{\overset{|}{C}}}}\text{—}COOC_2H_5 + H_2O$$

Ethyl ester of AA.

(c) *Reaction with amines.* Forms amides.

$$R\text{—}\underset{NH_2}{\underset{|}{CH}}\text{—}COOH + HHN - R \rightarrow R\text{—}\underset{NH_2}{\underset{|}{CH}}\text{—}CO\text{—}NH\text{—}R + H_2O$$

9. ***Reaction involving NH2 groups***

(a) *Reaction with mineral acids* (Salt formation). When AA or proteins are treated with HCl the acid salts are formed.

(b) *Reaction with formaldehyde.* With formaldehyde, he hydroxy methyl derivatives are formed which are insoluble in water and resistant to the attack of microorganisms.

(c) *Reaction with Nitrous acid* (*Van Slyke reaction*). The amino acids react with HNO_2 to liberate N_2 gas and to produce a-hydroxy acid.

(d) *Reaction with fluro-dinitrobenzene* (*FDNB*) or *Sanger's reagent.* In mildly alkaline solution, 1-fluro - 2, 4-dinitrobenzene reacts with α-amino acids to produce yellow coloured derivative, DNB-amino acid.

10. *Reaction involving both COOH and NH_2 groups* (a) Reaction with triketohydrindene hydrate (*Ninhydrin reaction*). Ninhydrin in a powerful oxidizing agent causes oxidative decarboxylation of α-amino acids producing CO_2, NH_3 and an aldehyde. The reduced ninhydrin then reacts with the liberated NH_3 forming blue-coloured Ruheman's complex.

11. ***Reactions involving—R group or side chain***

(a) *Biuret test.* Compounds containing peptide bonds produce a purple colour when treats with Biuret reagent (.2% alkaline $CuSO_4$). A substance Biuret is formed. Dipeptides do not respond to this reaction.

(b) *Xenthoproteic test.* Yellow colour develops when proteins are boiled with conc. HNO_3 due to presence of benzene ring. This reaction is due to nitration of phenyl rings. This is the test for tyrosine, tryptophan, phenylalanine.

(c) *Million's test.* When proteins are heated with $HgNO_3$ in HNO_3, a red colour develops. This reaction is specific for tyrosine. Tryptophan also responds to this reaction.

(d) *Hoplin's—Cole test* (or *Glyoxylic acid test*). Violet ring develops on addition of Conc. H_2SO_4 at the junction of protein and glyoxylic acid solution. This test is specific for tryptophan.

(e) *Folin's test.* Blue colour develops with phosphomolybdo-tungstic acid in alkaline soln-due to phenol group. Specific test for tyrosine.

Biological Importance of Proteins

Proteins constitute a large part of the cell-structure and are present in all the tissue. Many proteins have special physiological functions.

1. *Membrane proteins.* The integral proteins include translocases, the peripheral proteins include cytochrome—C and monamine oxidase.
2. *Enzymes.* The enzymes are biocatalysts to influence the rats of a chemical reaction. All the enzymes are proteins.

3. *Hormones*. Some of the messangers of our body, the hromones, are proteins.
4. *Blood proteins*. 6 major plasma components are albumen, α-1 globulin, α-2 globulin, β-globulin, γ-globilin and fibrinogen. All are proteins.

CARBOHYDRATES

The class of substances, known as carbohydrates, is comprises of a large number of relatively heterogenous compounds. These are synthesized form carbon dioxide and water in chlorophyll bearing cells during the process of photosynthesis. They are especial constituents of plants (cellulose, starch etc.) but also occur and serve important functions in animals. They serve as the chief source of energy in the food of many animals. Their energy value is 4 cal per gram and this energy is provided to the various synthetic needs of a cell. In carbohydrate carbon, hydrogen and oxygen are generally found in 1:2:1 ratio.

Structure

Structurally the carbohydrate are the polyhydric alcohols of carbon possessing active aldehyde or ketonic groups, which on hydrolysis yield to such products.

CLASSIFICATION

The carbohydrates may be classified according to their complexity. They are classified into following groups:

1. Monosaccharides
2. Oligosaccharides
3. Polysaccharides.

Monosaccharides

Monosaccharides are those sugars which cannot be hydrolysed into a simple form. Their general formula is $(CH_2O)n$. The value of n ranges from 3-7. If the monosaccharide has an aldehyde group—CHO it is called an aldose e.g. glucose and if a keto group $> C = O$ is present it is called a ketose e.g. fructose.

Monosacchrides include trioses, tetroses, pentoses, hexoses, heptoses. This classification is based on the basis of carbon atoms present in their molecules. The recent trend in the classification of monosaccharides is a combination of both the system. The aldeosugars and ketosesugars are classified as below. The suffix ('- *ore*) denotes that the sugar belongs to the aldosugar grop and the suffix (*-ulose*) denotes the ketosugas.

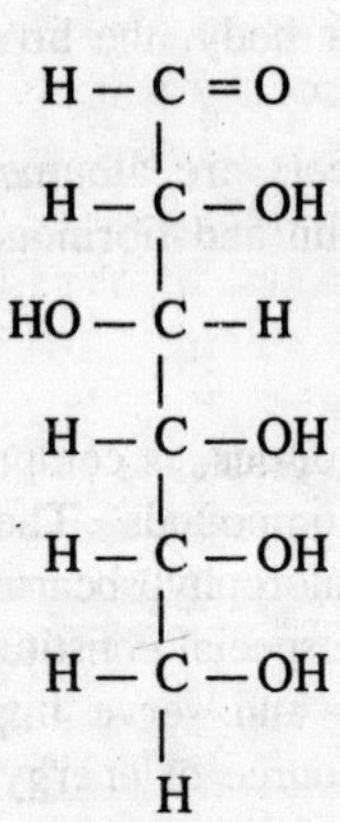

```
H — C = O
    |
H — C — OH
    |
HO — C — H
    |
H — C — OH
    |
H — C — OH
    |
H — C — OH
    |
    H
D-glucose
```

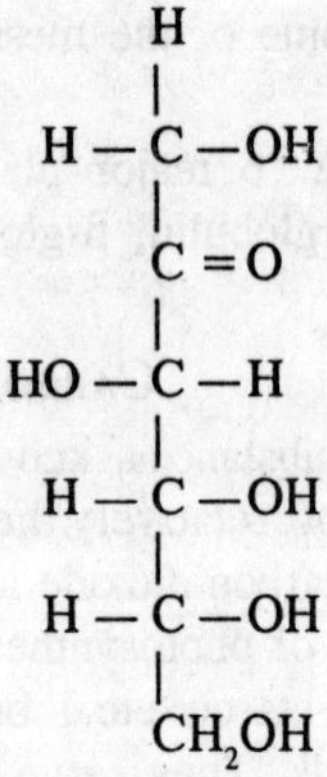

```
      H
      |
H — C — OH
      |
    C = O
      |
HO — C — H
      |
H — C — OH
      |
H — C — OH
      |
    CH2OH
D-fructose
```

CH$_2$OH, C, O, HO—C—H, H, OH, C, H—C—OH, H, C, H, H

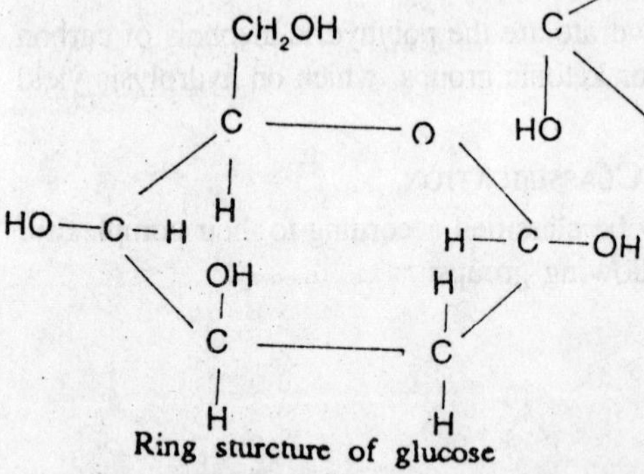

Ring sturcture of glucose

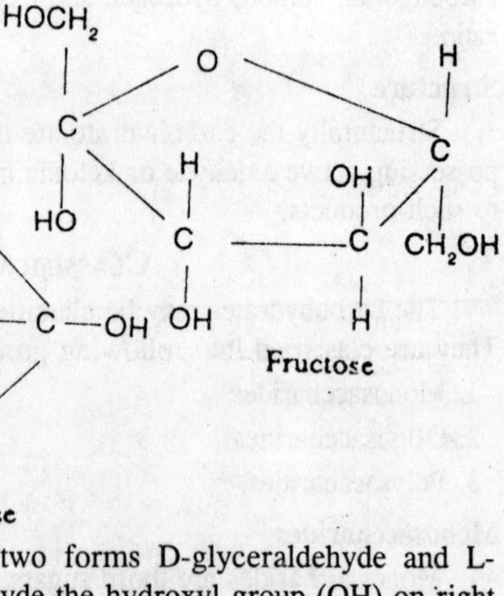

HOCH$_2$, O, H, C, C, H, OH, HO, C, C, CH$_2$OH, OH, H

Fructose

Glyceraldehyde occurs in two forms D-glyceraldehyde and L-glyceraldelyde. In D-glyceraldelyde the hydroxyl group (OH) on right side. In L-glyceraldelyde the position of the hydroxyl group is on the left on the asymmetric carbon atom.

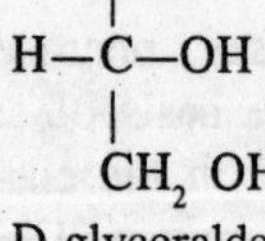

```
   CHO                 CHO
    |                   |
H—C—OH             OH—C—H
    |                   |
  CH2 OH              CH2 OH
D-glyceraldehyde    L-glyceraldehyde
```

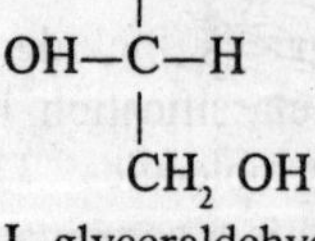

It has now been well-established that in the crystalline form, the monosaccharides containing 5 or more carbon atoms exist in the *tautomeric* ring forms. Even in solutions, these sugars mostly occur in the ring forms. The free sugars mostly contain six membered ring in which five members are carbon atoms and the remaining one member is an atom of oxygen. The six membered ring is known as *pyranose ring*. In adition, there is evidence of the existence of five membered rign in the sugar in which four members are carbon atoms and the remaining one member is the atom of oxygen. The five membered rign is known as *furanose rang*. Trioses and tetroses do not possess ring structure. The free pentose sugars are largely found in pyranose forms but in glycosides and nucleic acids these exist in the pyranose form.

Monosaccharides		*Type*	*Name*
1. Triose	$(C_3H_6O_3)$	Aldose	Glyceraldehyde
		ketose	Dihydroxy acetone
		Aldose	Erythrose
2. Tetrose	$(C_4H_8O_4)$	ketose	Erythrulose
3. Pentose	$(C_5H_{10}O_2)$	Aldose	Ribose
		Ketose	Ribulose
4. Hexoses	$(C_6H_{12}O_6)$	Aldose	Glucose, glactose
		Ketose	Fructose
		Aldose	Persenlose
5. Heptose	$(C_7H_{14}O_7)$	Ketose	Sedoheptulose

α-Form (glucose) β-Form (glucose)

In the pyranose type of hexoses, carbon at 'I' position is asymmetrical. If the hydroxyl group at position 'I' is *cis* to the hydroxyl group at position 2, it is known as α-form and if *trans* it is known as β-form. The α-form can readily pass into β-form when the sugar is brought into solution.

Properties of Monosaccharides

(a) Physical

The monosaccharides are sweet testing, colourless solids. They are soluble in water, partially in alcohols and insoluble in other.

When a polarized light is passed through a solution of these carbohydrates, the plane of light is rotated to either right or left side.

They contain asymmetric carbon atom hence exist in different isometric forms. The degree of optical rotation may change due to interconversion of isomeres. Fresh solution of glucose gives an optical rotation of +112° which change to + 52.7° on standing. The change in optical rotation is called *mutarotation.*

(b) Chemical

1. *Oxidation.* They can easily be oxidized by oxidizing agents. Glucose yields gluconic acid after oxidation with Tollen's reagent (ammonical Ag_2 O) or Fehlings solution (alkaline $CuSO_4$). The reduction of Tollen's reagent yields silver as polishing on the surface of tube whereas with the Fehling solution red ppt are obtained.

   When strong oxidizing agent is used like conc HNO_3, gluconic acid, is ultimately oxidizes to discarboxylic saccharic acid.

   Glucuronic acid is obtained in animal body on slow oxidation of glucose. Glucuronic acid combines with hormones. It is a major component of hyaluronic acid, heparin, mucoitin sulphate and chondriotine which are found in blood, skin and cartilage.
2. *Reduction.* Free aldehyde and ketone groups of mono-saccharides are reduced to alcoholic hydroxy groups by sodium—mercury amalgam and water. D-glucose after reduction yields a mixture of sorbitol and mannitol.
3. *Condensation.* Aldehyde groups of monosacharides condense with primary amines to form Schiff's base, when treated with hydroxylamine, glucose forms glucose oxime.
4. *Esterification.* The hydroxyl groups of alcohols in the carbohydrate may be converted to esters by treating with the appropriate acetylating agents. When D-glucose is treated with acetic anhydrine in the presence of pyridine, penta-acetyl glucose is formed.
5. *Methylation.* Methylating agents such as Ag_2O, CH_3OH react with monosaccharide to yield glycoside.
6. *Fermentation.* Monosaccharides such as glucose and fructose undergo alcholic fermentation by micro-organisms such as yeasts and produce ethanol and carbon dioxide.

Oligosaccharides

The oligosaccharides are those carbohydrates which on hydrolysis give two to five simple monosaccharide molecules. The oligosaccharides are composed of two to five monosaccharide units. During union of monosaccharde units water molecule is eleminated and the units are linked through an oxygen bridge. It is a glycosidic linkage. Their general formula is $(CH_2O)_{n-1}$.

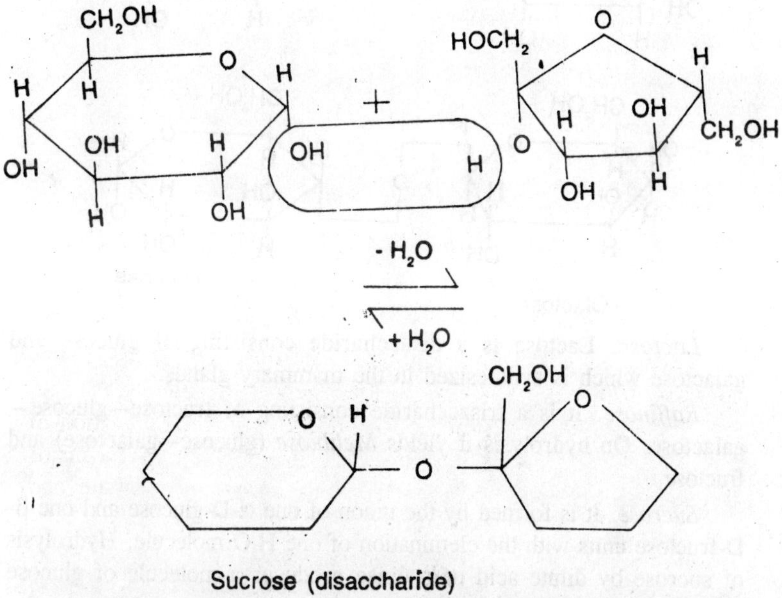
CH₂OH
O
H
H
H
OH
H
OH
OH
H
OH
+
HOCH₂
O
H
OH
H
CH₂OH
O
H
OH
H
- H₂O
+ H₂O
CH₂OH
O
O H
O

Sucrose (disaccharide)

The disaccharides have been classified into two groups namely, reducing and nonreducing. The reducing disaccharides are maltose, lactose, cellobiose, gentibiose and the non-reducing are sucrose and trehalose.

Cellobiose. Incomplete hydrolysis of cellulose given cellobiose. It is composed of two molecules of glucose linked by β-1, 4 glucosidic bond.

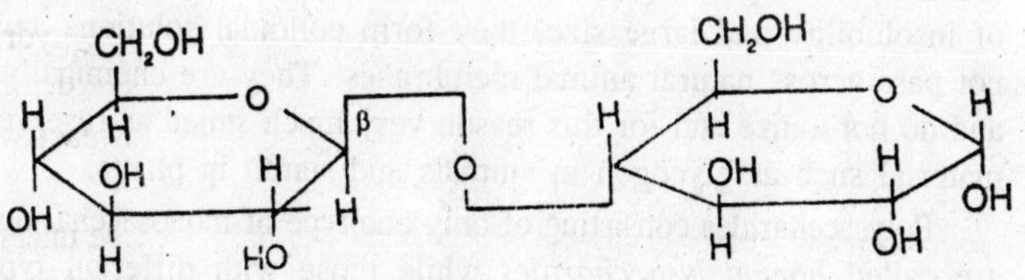
CH₂OH
O
H
H
OH
H
OH
H
HO
H
β
O
CH₂OH
O
H
H
H
OH
H
OH
H
OH

Maltose. It is composed of two units of D-glucose joined together through their 1 and 4 carbon atoms. It is obtained as a hydrolytic product by the action of amylase on strach

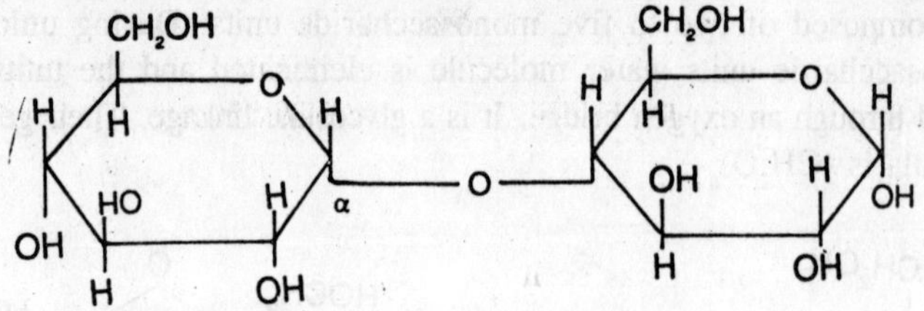

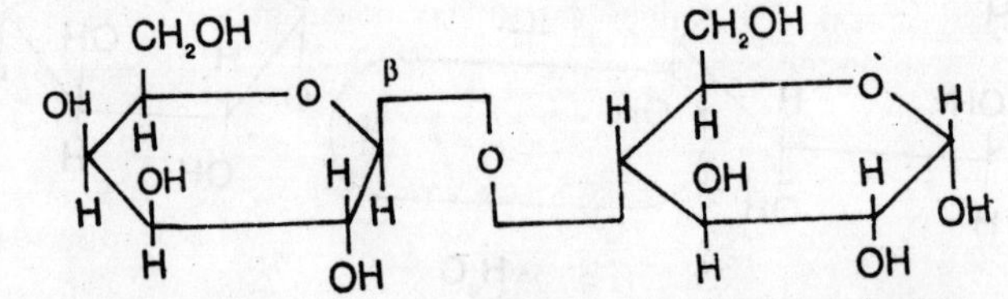

Glactose

Glucose

Lactose. Lactose is a disaccharide consisting of glucose and galactose which is synthesized in the mammary glands.

Raffinose. It is a trisaccharide consisting of fructose—glucose—galactose. On hydrolysis it yields *Melibiose* (glucose—galactose) and fructose.

Sucrose. It is formed by the union of one α-D-glucose and one β-D-fructose units with the elemination of one H_2O molecule. Hydrolysis of sucrose by dilute acid or sucrase produces a molecule of glucose and a molecule of fructose.

Polysaccharides

Polysaccharides may be regarded as carbohydrates formed by linking of a number of monosaccharides by glucosidic linkages. A carbohydrate having minimally 6 or more monosaccharide units may be regarded as polysaccharides. Their general formula is $(C_6H_{10}O_5)n$. They are tasteless, colourless amorphous powders which are insoluble in water. because of insolubility and large size, they form colloidal solutions and will not pass across natural animal membranes. They are chemically inert and do not ionize and for this reason very much suited as reserve food material such as glycogen in animals and starch in plants.

Polysaccharides consisting of only one type of monosaccharide units are called *homopolysaccharides* while those with different types of monosaccharde derivatives are *heteropolysaccharides*.

I. Homopolysaccchardes

1. *Glycogen.* It forms the carbohydrate reserve of the animal tissues and it is mainly stored in the muscles and liver. Fungi and yeast do also contain glycogen. It is a branched chain polymer having 6000—30,000 glucose units.
2. *Starch.* It is the most important food source of carbohydrate and is found in potatoes, rice, cereals etc. It is hydrolysed in gut yielding dextrins and maltose and eventually glucose.

Treatment of starch with hot water dissolves amylose, while amylopectine remains as such:

Amylose. Contains about 200-500 glucose units which are arranged in the form of a straight chain. The molecular wt. is about 150,000. It gives intense blue colour with iodine.

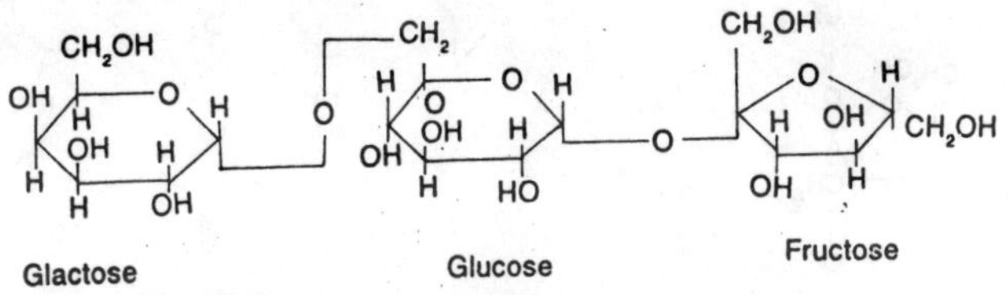

Glactose Glucose Fructose

Raffinose

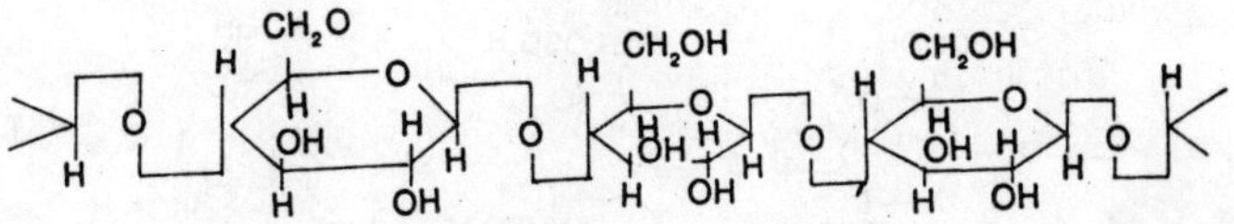

Cellulose

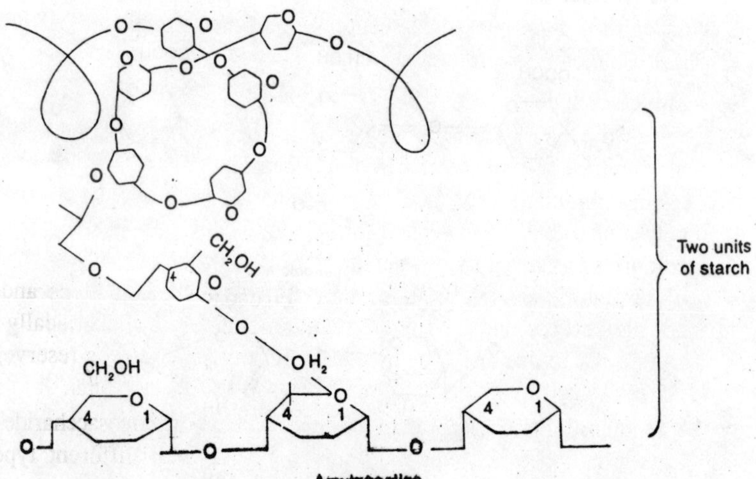

Two units of starch

Amylopectine

Amylopectin. It has about 1000 glucose resideus and its molecular wt. is about 200,000-1,00,000. It has a branched structure.

3. *Agar*. It is a galactan consisting of both D and L glactose. It is used as a bacteriological culture medium.
4. *Pectins*. These are abundant in fruits, particularly in the rim of citrus fruits like orange and lemons. They contain arabinose, glactose and galacturonic acid.
5. *Xylan*. In addition to cellulose all plants contain xylan. It is a hemicellulose and consists of D-xylose.

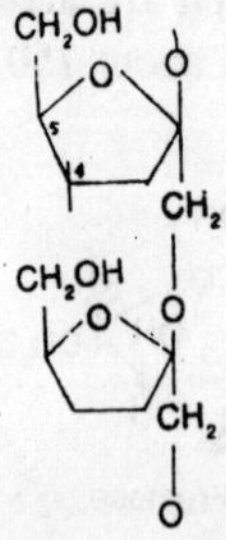

CH_2OH CH_2 CH_2OH CH_2 O

Inulin

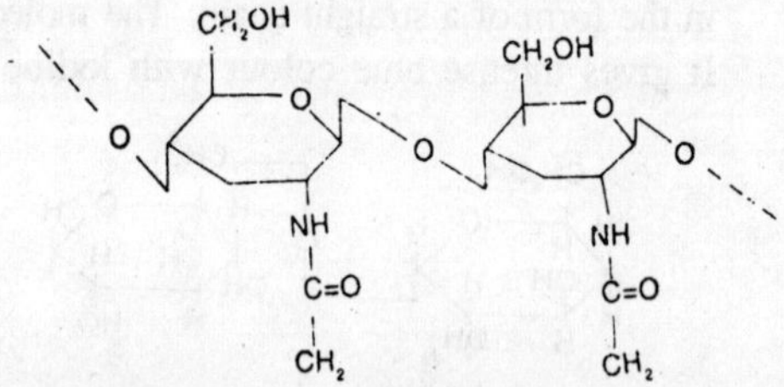

CH_2OH CH_2OH NH C=O CH_2 NH C=O CH_2

Chitiu

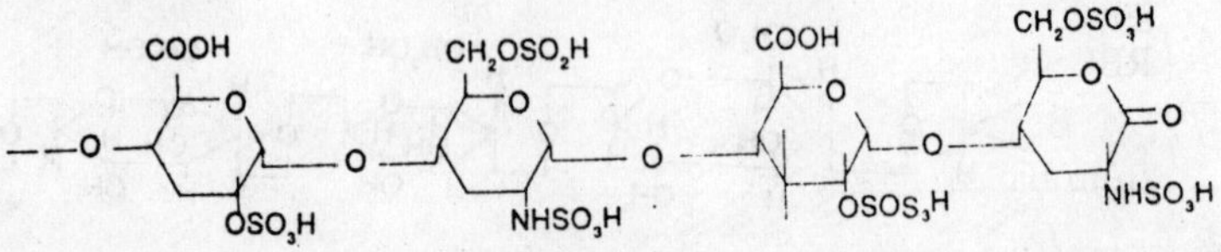

COOH OSO_3H CH_2OSO_2H $NHSO_3H$ COOH $OSOS_3H$ CH_2OSO_3H $NHSO_3H$

Heparin

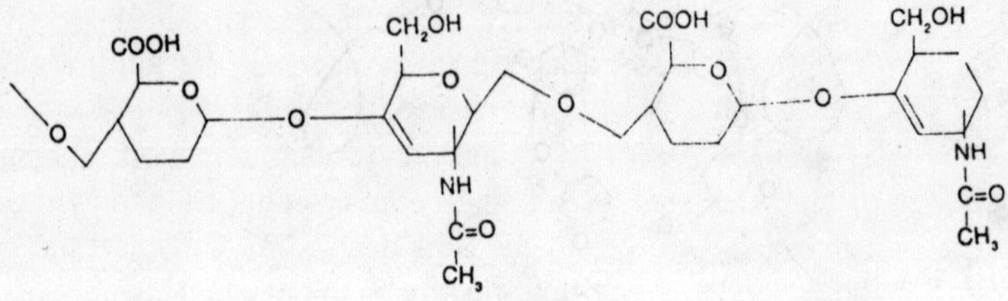

COOH CH_2OH NH C=O CH_3 COOH CH_2OH NH C=O CH_3

Hyaluronic acid

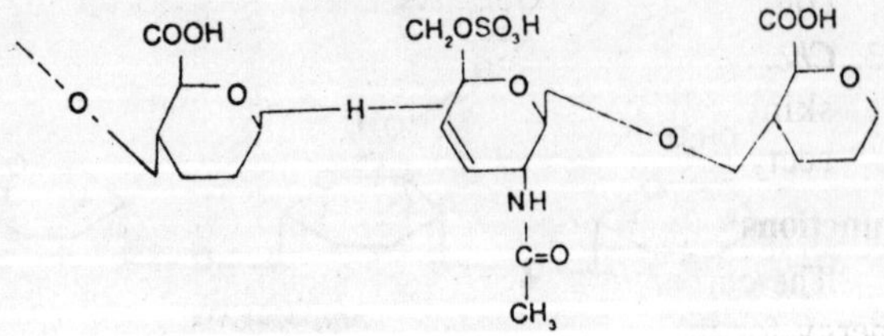

COOH CH_2OSO_3H NH C=O CH_3 COOH

chondriotin sulfate

6. *Inulin.* It is a starch found in the tubers and roots of *Dahlia*, *Dandelions* etc. On hydrolysis it yields fructose. M. Wt. is about 5,000 and about 30-35 fructose units per mole are present.
7. *Dextrins.* They are the intermediate products formed during hydrolysis of starch to the glucose.
8. *Cellulose.* The cell wall of plants is formed by cellulose. The cellulose consists of a lenier chain of glucose units ranging 900-2000. It does not occur in the animal tissue and it cannot be utilized by man for energy production purposes. Herbivorous animals can, however, utilize cellulose with the help of microorganisms as they can digest cellulose.
9. *Chitin.* It is an important structural polysaccharide in invertebrates particularly in arthropods. Structurally, it consists of N-acetyl-D glucosamine units.

II. Heteropolysaccharides

A. ***Glycoproteins.*** These are protein-polysaccharide compounds occuring in tissue, particularly in mucus secretion. Examples are ovalbumin, fibrinogen, γ-globulin of serum, human chorionic gonadotropins, luteinizing hormones and the blood group substances of RBC (antigen).

B. ***Mucopolysaccharides.*** The most important mucopolysaccharides from biological point of view are:

1. *Heparin.* It is present in the liver, lungs and spleen having a molecular wt. about 20,000. The *exect* structure of heparin in unknown. It is found to be constituted by glucuronic acid, glucosamine and sulfuric acid. Sulfuric acid is linked to hydroxyl group of sugar derivatives as well as to amino group of glucosamine. Heparin prevents clotting of blood in the vessels.
2. *Hyaluronic acid.* It is a mucopolysaccharide present in connective tissue and acts as an intercellular connecting material. It is abundently present in the umbilical cord, vitreous fluid of eyes and in the synovial fluid present at the joints. This polysaccharide consists of N-acetyl glucosamine and glucuronic acid.
3. *Chondrotin sulfates.* They are found in the cartilage, adult bones, skin, cornea, tendons and heart valves. It consists of N-acetylgalactosamine and glucuronic acid.

Functions

The carbohydrates are very important biologically as a source of energy for the cell. The main source of energy is glucose. It the main form of crbohydrates which is transported form cell to cell by blood in animal and by sap in plants. Carbohydrates also help in the formation of cell wall in plants. Some carbohydrates may serve as the prosthetic group of certain proteins. Glycogen and starch are major storage materials in animals and plants respectively.

LIPIDS

Term lipid includes fats and fat-like substances. Strucrurally, the different substances of this group may not be similar to fats, but all are soluble in fat-solvents like alcohols, ether, chloroform, carbon tetrachloride, acetone etc. They are insoluble in water. Lipids, like carbohydrates also, contain carbon, oxygen and hydrogen. The carbon and hydrogen are present in larger quantity than oxygen. Some lipids have phosphorous and nitrogen. The lipids in the cell may serve as condensed reserve of energy as well as form the membranous structures.

CLASSIFICATION OF LIPIDS

Lipids are classified into:

I. Simple Lipids

II. Compound Lipids

III. Derived Lipids

I. Simple Lipids

These are esters of fatty acids with various alcohols.

1. Natural Fats

These are the esters of fatty acids with glycerol. The chemical term for a natural fat is *triglycerides*. The triglycerides are formed by the combination of three molecules of fatty acids joined to one molecule of glycerol. As already stated triglycerides are neurtral fats, neutral because the acid ions are neutralized during their unification with glycerol; fats because the bulk of the molecule is devoid of electro-negative elements which can unite with hydrogen to form water molecule.

$$
\begin{array}{lcccl}
CH_2\ OH & & HOOC-R_1 & & CH_2OCOR_1 \\
| & & & & | \\
CHOH & + & HOOC-R_2 & \rightarrow & CHOCOR_2 \\
| & & & & | \\
CH_2OH & & HOOC-R_3 & & CH_2OCOR_3 \\
\text{Glycerol} & & \text{3 fatty acids} & & \text{Triglyceride} \\
\text{molecule} & & \text{molecules} & & +\ 3H_2O
\end{array}
$$

Triglycerides can exist in the solid or liquid form. It is based on the kind of fatty acid residues in its structure. The triglycerides which are blow 20°C called *oils* containing a large proportion of short-chained unsaturated fatty acids like *oleic, linoleic acid* etc. The triglycerides which are solid above 20°C are called *fats.* They contain long chained saturated fatty acids like *palmitic acid* and *stearic acid.*

2. *Fatty acids*

These are obtained by the hydrolysis of fats. These are monocarboxylic acids. The molecules of a fatty acid has a polar carboxyl group soluble in water and a non-polar hydrocarbon chain soluble only in fat solvents.

Table 2.3. Showing a few common fatty acids found in lipids.

Fatty acid		*Formula*
Saturated fatty acid	→	
Butyric acid	→	$CH_3(CH_2)_2$ COOH
Caproic acid	→	$CH_3(CH_2)_4$ COOH
Palmitic acid	→	CH_3 $(CH_2)_{14}$ COOH
Steraric acid	→	CH_3 $(CH_2)_{16}$ COOH
Unsaturated fatty acids		
Palmitoleic acid		CH_3 $(CH_2)_5$ (H=CH $(CH_2)_7$ COOH
Oleic acid		CH_3 $(CH_2)_7$ CH=CH $(CH_2)_7$ COOH
Linoleic acid		CH_3 $(CH_2)_4$ CH=CH CH_2 CH=CH $(CH_2)_7$ COOH
Linolenic acid		CH_3 CH_2 CH=CH CH_2 CH=CH CH_2 CH=CH $(CH_2)_7$ COOH

The fatty acids are classifed under three groups based on their degree of saturation and unsaturation.

(i) *Saturated fatty acids.* A saturated fatty acid contains as many hydrogen atoms as its carbon chain can hold. General formula for saturated fatty acid is R—COOH, where R is CH_3 (CH_2)n. n is varying from zero in acetic acid to 86 in mycolic acid. The most abunant saturated fatty acids in nature *palmitic* (C_{18}) and *Stearic* (C_{16}) *acids.*

(ii) *Unsaturated fatty acids.* They have one or more double bonds in their carbon chain. When there is a single double bond, results in the loss of 2 hydrogen atoms, such fatty acids are called *monous saturated fatty acids.* Their general formula is CnH_{2n}—1—COOH, examples are *palmitoletic acid* and *oleic acid.* When there ae 2,3,4 or more double bonds in the carbon chain with the consequent absence of 2,6,8 or more hydrogen atoms, such fatty acids are *polyunsaturated fatty acids*. The general formula is CnH_{2n}—2 COOH or CnH_{2n}—3 COOH. Examples are *linoleic acid* and *archidonic acids*.

Linoleic, linolenic and archidonic acid are often termed as *essential fatty acids*.

3. Waxes

Wexes are another class of simple lipis containing one molecule of fatty acid and one high molecular weight alcohol. The constituent acids and alcohols have usually 24-36 carbon atoms. The waxes has high melting points. They are chemically inert as they do not have double bonds in their hydrocarbon chains and are highly insoluble in water. They serve as protective converings on leaf surface (plants). *Bees wax* is an ester of palmitic acid with myricyl alcohol ($C_{30}H_{61}OH$) and *spermaceti* from the sperm whale is an ester of palmitic acid with cetyl alcohol ($C_{16}H_{33}OH$).

II. Compound Lipids

These are esters of fatty acids and alcohol with additional compounds, such as phosphoric acid, sugars, proteins etc. These are classified as follows:

(i) Phospholipids

Lipids containing phosphorous are phospholipids. They also have nitrogen containing bases and other substituents. The phospholipids are abundant in brain and nervous tissues. The different types of phospholipids are:

(a) *Phosphatidic acids*. These are compounds consisting of glycerol, two fatty acids and a phosphate group.

```
 H2C—OOR
    |
RCOOCH       O
    |        ||
 H2—C———O—P—OH
             |
             O-
```

Here R_1 and R_2 represent the residues of the molecules of fatty acids. The phosphatidic acids do not found in any great quantity in tissues but these are important as in intermediate biosynthesis of triglycerides and other phospholipids.

(b) *Lecithins.* Lecithins are choline esters of phosphatidic acid. On hydrolysis they yield one molecule of glycerol, 2 molecules of fatty acids and one molecule of phosphoric acid to which a nitrogenous base choline is attached.

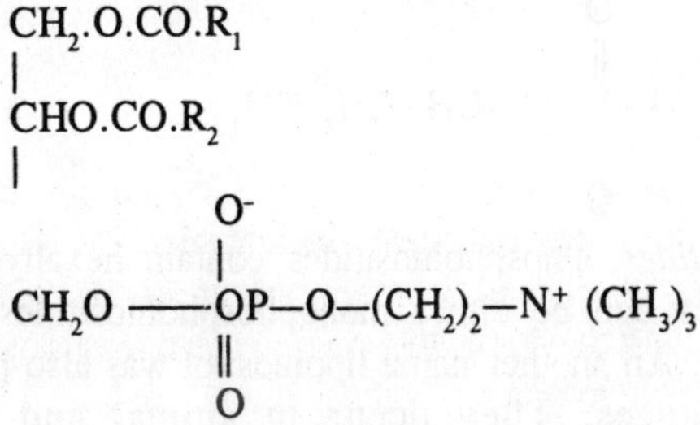

$$\begin{array}{l} CH_2.O.CO.R_1 \\ | \\ CHO.CO.R_2 \\ | \\ \qquad\qquad\quad O^- \\ \qquad\qquad\quad | \\ CH_2O \text{——} OP\text{—}O\text{—}(CH_2)_2\text{—}N^+ (CH_3)_3 \\ \qquad\qquad\quad \| \\ \qquad\qquad\quad O \end{array}$$

The fatty acids are palmitic, stearic, oleic, linolic and arachidonic acids. Lecithins are yellowish grey solids, soluble in ether and alcohols but insoluble in acetone. Lecithins get broken down by the enzyme lecithins to lysolecithin. This product (lysolecithin) has the ability to hemolyze the red blood corpuscles. Enzyme lecithinase occurs in the venom of snake cobra. and bee.

Lecithin is an important constituent of lipoproteins. Egg-yolk is a rich source of lecithin which also has an important role in fat metabolism in the liver.

(c) *Cephalins.* They resemble lecithins in most properties. The fundamental difference between the lecithins and cephalins is the nature of nitrogenous base. Cephalins contain ethanolamine, colamine and sometimes serine in place of choline.

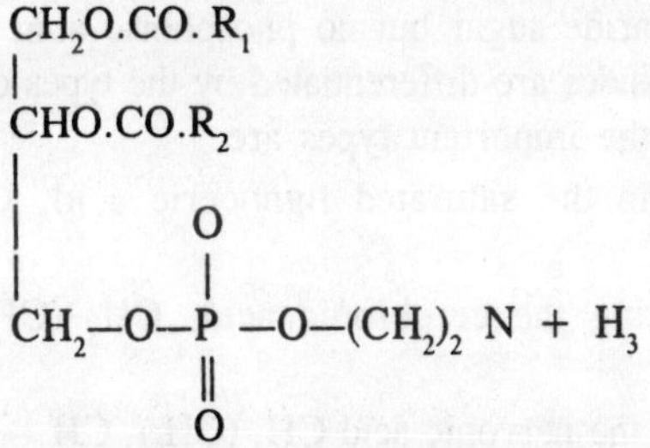

$$\begin{array}{l} CH_2O.CO.\ R_1 \\ | \\ CHO.CO.R_2 \\ | \qquad\quad O \\ | \qquad\quad | \\ CH_2\text{—}O\text{—}P\ \text{—}O\text{—}(CH_2)_2\ N + H_3 \\ \qquad\qquad \| \\ \qquad\qquad O \end{array}$$

(d) *Plasmalogens.* These phospholipids are abundant in brian and muscle. They resemble lecithins and cephalins in structure but possess an aldehyde group in place of one of the fatty acids in typical phospholipid molecule.

$$
\begin{array}{l}
CH_2OCH = CH_2R \\
| \\
CHOOCR_2 \\
| \qquad\quad O \\
| \qquad\quad \| \\
CH_2\ O{-}P{-}O{-}CH_2\ CH_2\ NH_3 \\
\qquad\quad | \\
\qquad\quad O
\end{array}
$$

(e) *Phosphoinositides.* Phosphoinositides contain hexahydric alcohol inositol. They can be either monophosphoinositides or diphosphoinositides. An another name lipoinositol was also proposed for these substances. They occur in animal and plants but diphosphoinositides have been reported from brain tissue only.

(f) *Phosphosphingosides.* Phosphosphingosides have been found to occur in high concentration in brain, nerves, lungs and spleen tissue. These contain a nitrogenous base sphingosine or its derivative which remains attached to long-chain fatty acid by its amino group. On hydrolysis they yield fatty acids, phosphoric acid, choline and sphingosine.

(ii) Glycolipids

The glycolipids occurs in brain tissue and in the myelin sheath of nerves, lungs, kidneys, spleen, liver, retina, egg-yolk etc. The glycolipids include two major catagorics, cerebrosides and gangliosides.

(a) *Cerebrosides.* They occur in the brain and myelin sheath of nerves. they are based on sphinge mine and have in addition a fatty acid and a monosaccharide sugar but no phosphoric acid or glycerol. Individual cerebrosides are differentiated by the types of fatty acids in the molecule. the important types are:

Kerasin containing the saturated lignoceric acid, $CH_3\ (CH_2)_{22}$ COOH.

Cerebron containing the cerebronic acid, $CH_3\ (CH_2)_{21}$ CHOH COOH.

Nervon containing the nervonic acid $CH_3\ (CH_2)_7\ CH = CH\ (CH_2)_{13}$ COOH.

Oxynervone containing hydroxy derivatives of nervonic acid $CH_3\ (CH_2)_7\ CH{=}CH\ (CH_2)_{12}$ CHOH COOH.

(b) *Gangliosides.* These contain N-acetylneuraminic acid (sialic acid), fatty acids, sphingosine and 3 molecules of hexose.

III. Derived Lipids

These lipids include hydrolytic products of lipids as well as other lipid like compounds like sterols, carotenoids, hydrocarbous etc.

The *sterols* are solid wax-like substances chemically, they are alcohols and occur either as such or as esters of fatty acids. They are highly soluble in fat solvents. They all contain a cyclopentano-phenanthrene (sterane) nucleus (made up of three cyclohexane rings, in the phenanthrane type of arrangement and a terminal cyclopentane ring).

Cholesterol is animal origin occuring in bile, brain, spinal cord etc. It is obtained form human gall bladder stones which are deposited in the bile duct.

Ergosterol. It is similar to 7-dehydrocholesterol but differs in the side chain. It can be converted into vitamin D on exposure to the ultra violet light and, therefore, also called provitamin D.

Carotenoids. These are also included in lipids because of their solubility in fat solvent. Carontenes consists of carbon and hydrogen only where as xanthophyll contain oxygen in addition. Vitamin A is derived from carotene. *Estrogens, progesterone, testosterone* and *anderosterone* are steroidal hormoens.

Prostaglandins

Prostaglandins are derivatives of fatty acids. They were first discovered in human seminal fluid secreted by prostate gland, hence the name. It has been shown now that prostaglandins are synthesized and released by many other tissues such as kidneys, testis, placenta, lungs, liver, uterus, gastrointestinal tract, brain and heart.

The prostaglandins are 20 carbon fatty acids including a five membered ring in their molecular structure. Different prostaglandins differ with one another in the number and position of double bonds and hydroxyl group substituents.

Prostaglandins PGE, PGE_1, PGE_3, PGF_{1a}, PGF_{2a} and PGF_{3a} are considered as *primary prostaglandins.* Most of their action appear to be binding of hormones to membrane.

Recently, new type of prostaglandin has been isolated from human seminal fluid which has been designated as PGX^2.

Amino Acids

There is a common plan of construction for the thousands kinds of proteins in living systems. The 20 kinds of naturally occurring amino acid monomers are strung together in unbranched, linear polymer chains

Table 2.4. Showing Elements that Occur in Protoplasm

S.No.	*Name of elements*	*Approximate percentage*	*Function*
Major Elements			
1.	Carbon	18	Forms backbone of all organic molecules.
2.	Hydrogen	10	Present in most organic com-pounds and major components of water.
3.	Oxygen	65	Cellular respiration; in organic compound and component of water.
4.	Nitrogen	03	Major components of all the amino acids, proteins and nucleic acids.
Tracer Elements			
5.	Sodium	0.2	For water balance, conduction of nerve impulse.
6.	Potssium	0.5	Conduction of nerve impulse, muscular contraction.
7.	Phosphorus	1.0	Nucleic acids formation; bones formation and in energy transfer.
8.	Magnesium	0.1	Constituent of certain enzymes (ATPase).
9.	Calcium	1.5	Blood clotting, muscle contraction, bones and teeth formation.
10.	Sulfur	0.3	Constituent of most proteins.
11.	Chloride	0.1	Negative ion of interstitial fluid.
12.	Iron	0.01	Component of haemoglobin and certain enzymes.

of proteins. These are the 20 amino acids specified in the genetic code that is universal to all organisms. Some other kins of amino acids are also found in cells, but they are either degradation products or residues that have been modified form oen of the 20 commonly occurring amino acids after this latter has been inserted into the polymer chain. Hydroxyproline is a major amino acid constituent of collagen in connective tissue, but proline residues are initially included in the protein and are converted to hydroxyproline after polymerization. Hydroxyproline is not one of the encoded acid, but proline is. Many proteins contain fewer than 20 kinds of amino acids. The relative proportions and the absolute number of the amino acid repertory vary from one protein to another, as a reflection of the specific in formation in genes, which are the blue prints for protein construction.

Nucleotides and Nucleic Acids

Nucleotides are involved in at least two major cellular functions: (1) they are monomeric units from which DNA and RNA polymers are constructed, and (2) they act as agents in certain energy-transferring reactions during metabolism. A mononucleotide is made up of one nitrogen-containing organic base, one pentose sugar, and one phosphate residue derivative from phosphoric acid. When there is no phosphate group, the sugar-base combination is called a *nucleoside* (Table 2.5). For this reason, *nucleotides* (phosphate-sugar-base) are also called nucleoside phosphates. Nucleoside mono-, di-, and tri- phosphates contain one, two or three phosphate groups respectively.

Table 2.5. Showing nomenclature of nucleic acids and their constituent units.

Base	*Nucleoside*	*Nucleotide*	*Nucleic Acid*
Purines			
Adenine (A)	Adenosine	Adenylic acid	RNA
	Deoxyadenosine	Deoxyadenylic acid	DNA
Guamine (G)	Guanosine	Guanylic acid	RNA
	Deoxyguanosine	Deoxyadenylic acid	DNA
Pyrimidines			
Cytosine (C)	Cytidine	Cytidylic acid	RNA
	Deoxycytidine	Deoxyadenylic acid	DNA
Thymine (T)	Thymidine	Thymidylic acid	DNA
Uracil (V)	Uridine	Uridylic acid	RNA

The nitrogenous bases commonly found in nucleic acids and their nucleotide building blocks are derivatives of purine and pyrimidine. The commonly occuring purines *adenine* and *guanine* are found in both DNA and RNA, as in the pyrimidine compound *cytosine,* the second kind of pyrimidine in DNA is *thymine,* while its demethylated form, *uracil,* occurs in RNA.

Since each kind of nucleic acid contains one unique pyrimidine, it is convenient to study synthesis and activity of DNA or RNA using isotopically-labelled precursors containing one or the other of these bases. Usually the nucleosides uridine or thymidine, or other nucleotide forms, are added to the biological system under study.

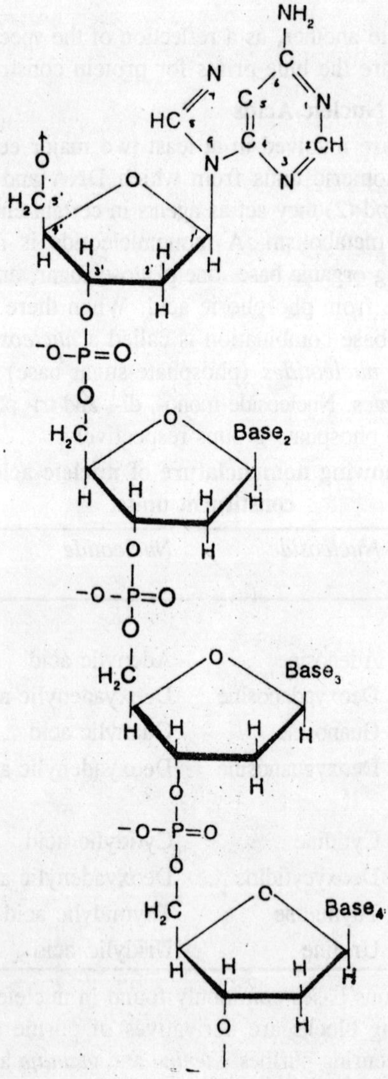

Fig. 2.6. The molecular organization of a backbone of DNA showing phosphodiester and glycosidic bonds.

The only difference in the pentose sugars of nucleotides is the presence of a hydroxyl group at carbon atom 2 of D-ribose in RNA, but a hydrogen at carbon-2 of 2-deoxy-D-ribose in DNA monomers and polymers. This seemingly simple difference is partly responsible for profound differences in stabilities, pairing potential, and functions of DNA and RNA.

Polynucleotides of both DNA and RNA varieties are built from mononucleotides that are linked convalently via phosphodiester bridges between the 3' position of one unit and the 5' position of hte next. Since there is no restriction on the ventrical sequence of adjacent mononucleotides in either DNA or RNA, a considerable variety of molecules is possible even though only 4 kinds of nucleotide monomers (one for each of the four kinds of bases in the combination with sugar and phosphate) are used in polymer construction. The theoretical variety is calcualted as 4", where 4 is the number of different kinds of nucleotides, and *n* is the number of monomers in the polymer. For a

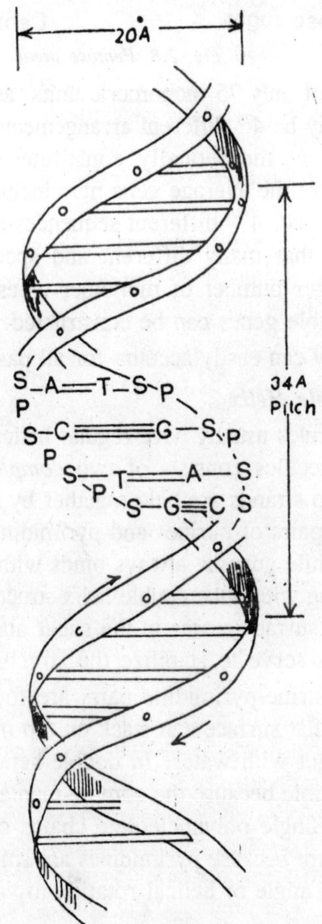

Fig. 2.7. DNA and its paired bases.

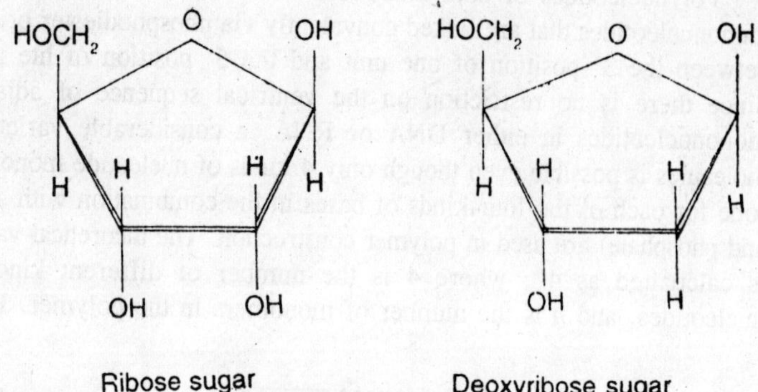

Fig. 2.8. Pentose sugar.

molecule made of only 75 monomeric units, as in some of the smallest RNAs, there may be 4^{75} different arrangements of the constituent units. Each arrangement theoretically constitutes a moleule of different specificity. Where the average gene may include about 500 nucleotides in a DNA sequence, 4^{500} different sequences are theoretically possible and, therefore, that many different and specific genes. Despite the apparently meager number of monomer types an astronomically high number of possible genes can be constructed.

Such variety can easily account for all past and present life forms.

The DNA Double Helix

DNA molecules usually have regular helical configurations because most DNA molecules consists of two *complementary polynucleotide strands*. The two strands are held together by *hydrogen bonds* between complementary pairs of purines and pyrimidines. Adenine always bind with thymine while guanine always binds with cytosine. This repeated hydrogen bonding within the double helix structure and bonding between virtually all the surface atoms in the sugar and phosphate groups with water molecules serve to stabilize the structure.

Since the purine-pyrimidine pairs are found in the center of the molecule, their flat surface can stack on top of each other and thereby limit their contact with water. In double-helical molecules, a regular structure is possible because the comple-mentary base pairs are exactly the same size. Single polynucleotide chains could not have a regular backbone structure because pyrimidines are smaller than purines, which would cause the angle of helical rotation to vary with the sequence of bases.

DNA double helix molecules are very stable at physiological temperatures because: (1) disruption of the double helix breaks hydrogen

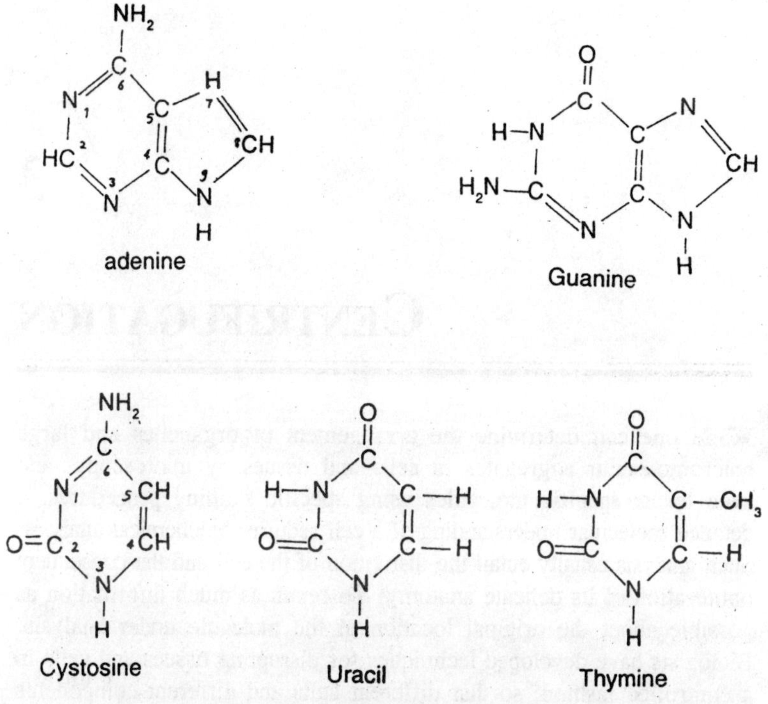

adenine

Guanine

Cystosine

Uracil

Thymine

Fig. 2.9. Various heterogenous bases.

bounds and brings hydrophobic purines and pyrimidines into contact with water, which is energetically unsatisfactory; and (2) there are many weak bonds within the DNA molecule, arranged so that most of them cannot break without many others breaking at the same time. Even though some hydrogen bonds may be broken by thermal motion, hydrogen bonds in the rest of the molecule remain intact and the molecule does not fall apart. In fact, when held together by more than ten nucleotide pairs the double helices are quite stable. At room temperature weak bonds is the stability of molecular shape, in proteins as well as in nucleic acids. At abnormally high temperatures there is more frequent breakage of weak bond, which become less stable as temperatures rise above physiological levels. Once a significant number of weak bonds have been broken, a protein or nucleic acid molecule usually loses its original form and changes to an inactive or denatured form.

3

CENTRIFUGATION

While one can determine the arrangement of organelles and large macromolecular aggregates in cells and tissues by microscopy, and even locate specific molecules using specific staining procedures, a detailed molecular understanding of a cell requires biochemical analysis. Such analysis usually entail the disruption of the cell and the consequent obliteration of its delicate anatomy. To retain as much information as possible about the original location of the molecule under analysis. Biologists have developed techniques for disrupting tissues and cells in a controlled fashion, so that different cells and different components of cells can be separated before biochemical analysis.

ISOLATION OF CELLS

Many types of differentiated cells are not readily obtained as a cultured cell line. In any case, it is usually cheaper and quicker to use cells isolated directly from an animal or plant for large-scale biochemical analysis. The disadvantage is that all tissues in a higher animal or plant contain a mixture of cell types, which must be separated before analysis. Suspensions of single cells are first prepared from the tissue by disrupting the extracellular matrix and intercellular junctions that hold the cells together. The best yields of viable dissociated cells are usually obtained from fetal or neonatal tissues. The procedures is to treat the tissues with proteolytic enzymes (such as trypsin and collagenase and agents that bind, or chelate, Ca^{2+} (such as ethylenediaminetetra-acetic acid, or EDTA) and then to dissociate them into single cells by gentle mechanical disruption.

Several approaches are used to separate the different cell types from a mixed cell suspension. One is to exploit the differences in the cells' physical properties. For example, large cells can be separated from small cells and dense cells from light cells by sedimentation or centrifugation; these techniques will be described when we discuss the separation of organelles and macromolecules, for which they were originally developed. Another approach is based on the fact that some cells adhere strongly to glass or plastic and therefore can be separated from cells that adhere less strongly.

An important refinement of this last technique depends on the specific binding properties of antibodies. Antibodies that bind specifically to the surface of only one cell type in a tissue can be coupled to various matrices—such as collagen, polysaccharide beads, or plastic—to form an "affinity surface" to which only cells recognized by the antibodies will adhere. The bound cells are then recovered by gentle shaking or, in the case of a digestible matrix (such as collagen), by degrading the matrix with enzymes such as collagenase).

The most sophisticated cell-separation technique involves labeling specific cells with antibodies coupled to a fluorescent dye and then separating the labeled cells from the unlabeled ones in an electronic fluorescence-activated cell sorter. Here, individual cells traveling in single file in a fine stream are assessed for their fluorescence by passing them through a laser beam. Slightly further downstream, tiny droplets, most containing either one or no cells, are formed by a vibrating nozzle. The droplets containing a single cell are automatically given a positive or a negative charge at the moment of formation, depending on whether they contain a fluorescent cell; they are then deflected by a strong electric field into an appropriate container. Occasional clumps of cells, detected by their increased light scattering, are left uncharged and are discarded into a waste container. Such machines can select 1 cell in 1000 and sort about 5000 cells each second.

Separation of Organelles and Macromolecules

The cells in a purified population can be disrupted in various ways: by osmotic shock, by ultrasonic vibration, by forcing the cells through a small orifice, or by grinding them up. These procedures break many of the membranes of the cell (including the plasma membrane and membranes of the endoplasmic reticulum and Golgi apparatus) into fragments that immediately reseal to form small, closed vesicles. But, if carefully applied, the disruption procedures leave organelles such as nuclei, mitochondria, lysosmoes, and peroxisomes intact.

The population of cells is thereby reduced to a soluble extract containing a thick suspension of membrane-bounded particles, each with a distinctive size, charge, and density,. Provided that the homogenization medium has been carefully chosen (this requires extensive trial and error for each organelle), the various particles retain most of the biochemical properties of the original organelles in the intact cell.

Separating the various components in this mixture became possible only after the commercial development in the early 1940s of an instrument known as the *preparative ultracentrifuge*, in which extracts of broken cells are rotated at high speeds. At a relatively low speed, large components, such as nuclei and unbroken cells, sediment rapidly and form a pellet at the bottom of the centrifuge tube; at a slightly higher speed, a pellet of mitochrondria is deposited; and at even higher speeds and longer periods of centrifugation, first the small, closed vesicles and then the ribosomes can be collected. All of these functions are impure, but resuspending the pellet and repeating the centrifugation procedure several times removes many of their contaminants.

A finger degree of separation can be achieved by layering the cell homogenate as a narrow band on top of a salt solution in a centrifuge tube. To stablize the sedimenting component against convective mixing, the salt solution beneath the band contains an increasingly dense solution of an inert, highly soluble material such as sucrose (a density gradient). Under these conditions, the different fractions sediment at different rates, forming distinct bands that can be individually collected. The rate at which each component sediments depends on its size and shape and is normally expressed as its sedimentation coefficient or s value (Table 3.1). Present-day ultracentrifuges rotate at speeds up to 80,000 rpm and produce forces up to 500,000 times gravity. At these enormous forces, even relatively small macromolecules, such as tRNA molecules and simple enzymes, separate from one another on the basis of their size.

Table 3.1. Some Typical Sedimentation Coeffiicients

Particle or Molecule	*Sedimentation Coefficient*
Lysosome	9400S
Tobacco mosaic virus	198S
Ribosome	80S
Ribosomal RNA molecule	28S
tRNA molecule	4S
Hemoglobin molecule	4.5S

The ultracentrifuge is also used to separate cellular components on the basis of their buoyant density rather than their size. In this case, the sample is sedimented through a steep gradient that contains a very high concentration of sucrose or cesium chloride. The cellular components move down the gradient until they reach a position that is equal to their own density, and at this point they float and can move no further. This method can be so sensitive that it is capable of resolving macromolecules that have incorporated heavy sedimentation coefficient (s), in units of seconds, are given by $(dx/dt)/w^{2x}$, where x is the distance from the center of rotation in centimeters, dx/dt is the speed of sedimentation in centimeters per second, and w is the angular rotation of the centifuge rotor in radians per second.

Because such coefficients are extremely small numbers, they are normally expressed in Swedberg units (S), where $1S = 1 \times 10^{-13}$ sec. isotopes, such as ^{13}C or ^{15}N, from the normal unlabeled species. In fact, the cesium chloride method was developed in 1957 to separate the labeled and unlabeled DNA produced after exposing a growing population of bacteria to nucleotide precursors containing ^{15}N; this classic experiment provided direct evidence for the semi-conservative replication of DNA.

Table 3.2. Major Events in the Development of the Ultracentrifuge and the Preparation of Cell-free Extracts

1897	*Buchner* showed that cell-free extracts of yeast can ferment sugars to form carbon dioxide and ethanol, laying the foundations of enzymology.
1926	*Svedberg* developed the first analytical ultracentrifuge and used it to estimate the molecular weight of hemoglobin as 68,000.
1935	*Pickels* and *Beams* introduced several new features of centrifuge design that led to its use as a preparative instrument.
1938	*Behrens* employed differential centrifugation to separate nuclei and cytoplasm from liver cells, a technique further developed for the fractionation of cell organelles by *Claude*, *Brachet*, *Hogeboom*, and others in the 1940s and early 1950s.
1949	*Szent-Gyorgyi* showed that isolated myofibrils from skeletal muscle cells contract on the addition of ATP. In 1955, a similar cell-free system was developed for ciliary beating by *Hofmann-Berling*.
1951	*Brake* used density gradient centrifugation in sucrose solutions to purify a plant virus.

1953 *de Duve* isolated lysosomes and, later, peroxisomes by centrifugation.

1954 *Zamecnik* and colleagues developed the first cell-free system to carry out protein synthesis. This was followed by a decade of intense research activity, during which the genetic code was elucidated.

1957 *Meselson*, *Stahl*, and *Vinograd* developed density gradient centrifugation in cesium chloride solutions for separating nucleic acids.

Centrifugation

Centrifuges are designed to accelerate sedimentation by utilizing centrifugal force. Various types of centrifuges are used in the clinical laboratory for separating suspended particles from a liquid in which the particles are not soluble. Liquids of differing specific gravities (density) may also be separated.

There are three general types of centrifuges: the horizontal head, the angle head, and the ultracentrifuge. Many variations of the horizontal and angle head units are found in the clinical laboratory. These include bench top and floor standing units, refrigerated units, and such special-purpose instruments as the microhematocrit, microsample, cytospin, and continuous-flow systems.

Principles of Centrifugation

In the clinical laboratory, the centrifuge function as a filtration or packing device. The development of large-batch centrifuges and continuous-flow systems which can sediment a precipitate from a large volume of solution in a short time have tended to replace the tedious process of filtration in most clinical and research laboratories. This development has not only speeded the process of separation but, when coupled with refrigeration, has reduced sample lability to a minimum.

General applications include the separation of serum or plasma from red blood cells, the separation of precipitated solids from the liquid phase of a mixture, or the separation of liquids of varying density. Special-purpose units include such applications as quantitative red blood cell packing for measuring hematocrit, automatic cell washing, and component preparation for blood banking. Ultracentrifuges employ very high speed, some with optical systems, to achieve difficult and precise quantitative separations of ultrasmall particles or macromolecules.

The horizontal-head shields or cups are in a vertical position when the centrifuge is at rest and assume a horizontal position when the

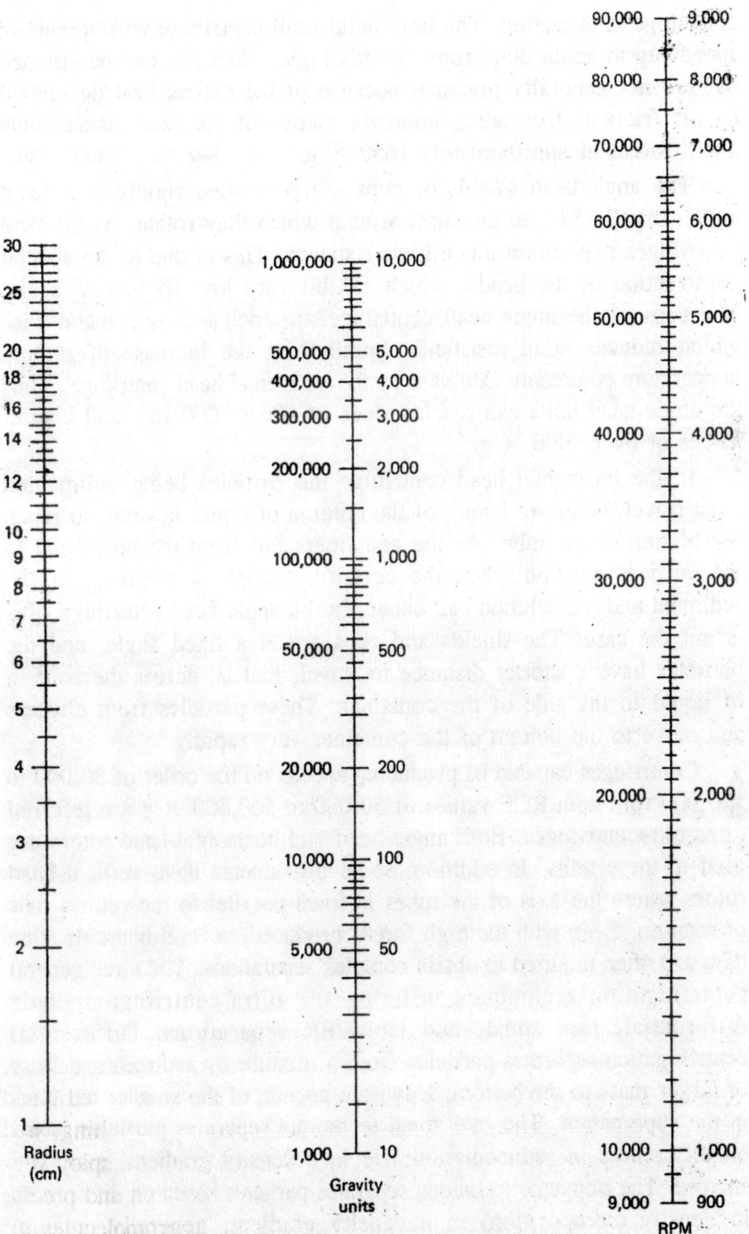

Fig. 3.1. Nomogram for the determination of the gravitational force of a centrifuge rotor. To determine g place the edge of a rular over the value for the radius of the rotor and run the edge over the speed of the rotor.

centrifuge is operating. The horizontal-head centrifuge will operate at speeds up to about 3000 rpm. Speeds higher than this can be attained but are not generally practical because of the excess heat developed by air friction. Depending upon the radius of the head, these units attain forces of approximately 1650 × g.

The angle-head shields or cups are positioned rigidly at a fixed angle, usually 52°, to the shaft around which they rotate. Angle-head centrifuges may attain much higher speeds. This is due to the special construction of the heads, which exhibit very low friction with air. The cups of the angle-head centrifuge are enclosed by a metal case which reduces wind resistance, minimizing the increase of sample temperature commonly exhibited by the horizontal-head centrifuge. Thus the angle-head units can reach speeds of about 7000 rpm and exhibit forces of over 9000 × g.

In the horizontal head centrifuge the particles being sedimented must travel the entire length of the column of liquid in order to reach the bottom of the tube. As the containers fall from the horizontal to the vertical position when the centrifuge stops, a remixing of the sediment and the solution can occur. In the angle-head centrifuge, this is not the case. The shields and cups are at a fixed angle, and the particles have a shorter distance to travel, that is, across the column of liquid to the side of the container. These particles from clusters and move to the bottom of the container very rapidly.

Centrifuges capable of producing speeds on the order of 30,000 to 100,000 rpm with RCF values of 50,000 to 500,000 × g are referred to as ultracentrifuges. Both angle-head and horizontal-head rotors are used in these units. In addition, some instruments have vertical-head rotors where the axis of the tubes is fixed parallel to the rotor's axis of rotation. Even with the high forces produced, several hours or even days are often required to obtain complex separations. The three general centrifugation techniques utilizing the ultra-centrifuge include differential, rate zonal, and isopycnic separations. Differential centrifugation separates particles from a mixture by sedimenting those of larger mass to the bottom, leaving a portion of the smaller particles in the supernatant. The rate zonal technique separates particles based on differences in sedimentation rate in a density gradient, generally sucrose. The isopycnic technique separates particles based on differences in density (composition) in a density gradient, generally, cesium chloride. The ultracentrifuge is significantly more complex than general laboratory centrifuges and has not as yet found wide-spread use in the

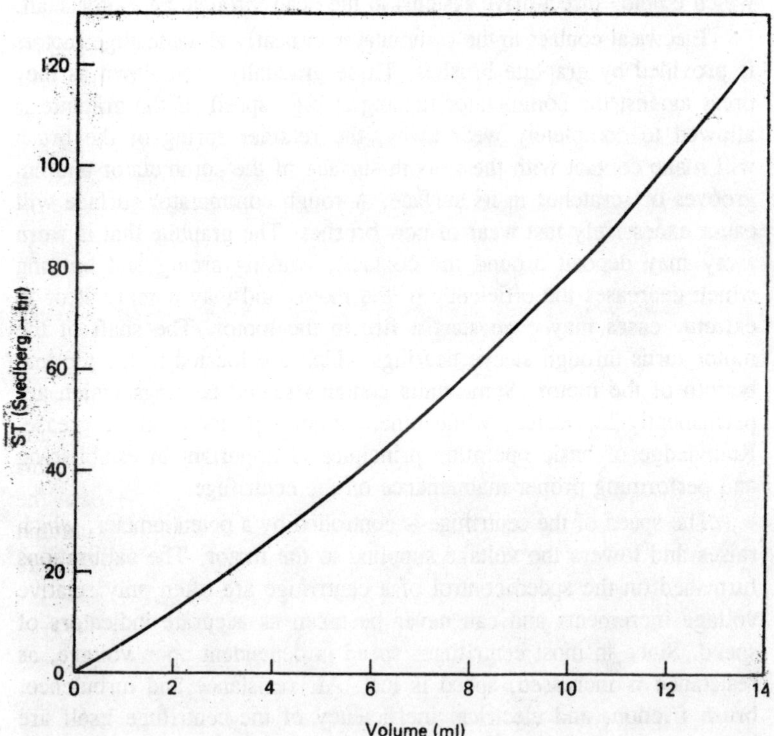

Fig. 3.2. Rotor characteristics for the rapid computation of $\overline{ST}$ values for given volumes traversed by particles in a Spino No. 40 rotor having a maximum speed of 40,000 rpm.

clinical laboratory; thus the information that follows does not pertain to the ultracentrifuge.

Component Parts

The parts of a common centrifuge include the chamber which encloses the internal parts, a cover with latch, the centrifuge head with shields or cups, the shaft and rotor on which the head turns, and the motor-drive assembly. Most centrifuges will include a power switch, braking device, speed control, timer, and possibly a tachometer.

Centrifuge motors are generally series-would dc motors that turn faster as voltage is increased. Occasionally, ac motors are utilized, and speed adjustment is achieved through stepwise reduction of the number of poles in the magnetic field. Both types are high-speed motors which employ direct-drive systems to the head through the motor shaft.

Electrical contact to the commutator in nearly all centrifuge motors is provided by graphite brushes. These gradually wear down as they press against the commutator turning at high speed. If the graphite is allowed to completely wear away, the retainer spring of the brush will make contact with the smooth surface of the commutator and cut grooves or scratches in its surface. A rough commutator surface will cause excessively fast wear of new brushes. The graphite that is worn away may deposit around the contacts, causing arcing and burning which decreases the efficiency of the motor and may damage it or in extreme cases may even start a fire in the motor. The shaft of the motor turns through sleeve bearings which are located at the top and bottom of the motor. Some units contains sealed bearings which are permanently lubricated, while others require periodic oil or grease. Knowledge of basic operating principles is important in establishing and performing proper maintenance on the centrifuge.

The speed of the centrifuge is controlled by a potentiometer, which raises and lowers the voltage supplied to the motor. The calibrations furnished on the speed control of a centrifuge are often only relative voltage increments and can never be taken as accurate indicators of speed. Since in most centrifuges speed is dependent upon voltage, as resistance is increased, speed is lost. Air resistance and turbulence, brush friction, and electrical inefficiency of the centrifuge itself are sources of resistance. These resistances can cause a centrifuge to operate at differing speeds on the same speed-control setting. Different accessories and varying states of repair will also result in the same speed-control problems. Calibration and periodic recalibration of centrifuge speed is extremely important.

Operating of centrifuge with the lid raised is dangerous and must be discouraged. In addition to being hazardous and reducing the centrifuge's speed through increased air resistance, this will also cause the revolving parts of the centrifuge to vibrate excessively, thereby causing extensive wear on the centrifuge and a remixing of the sedimented particles.

Tachometers are provided on many centrifuges to indicate the speed in rpm. A flexible shaft or cable attached to the motor spindle

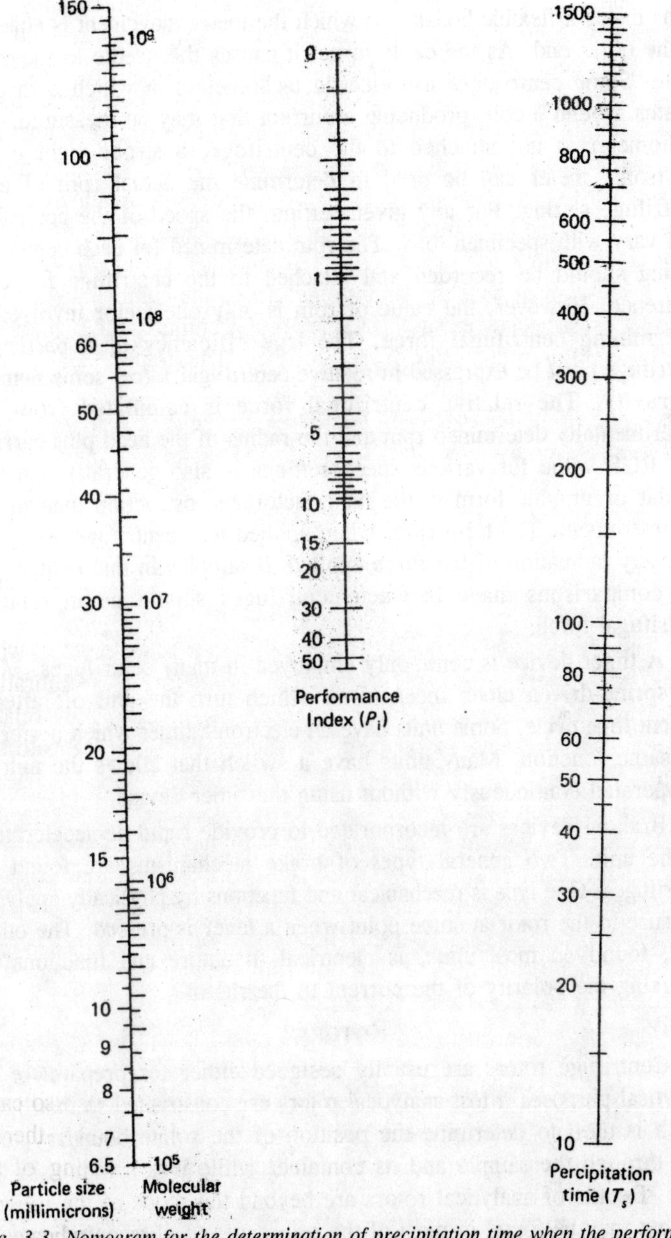

Fig. 3.3. Nomogram for the determination of precipitation time when the performance index of the rotor and the size of the sedimenting particles are known.

turns inside a flexible housing to which the meter movement is attached at the other end. As the cable turns, it causes the needle to move up scale. Some centrifuges use electric tachometers in which a magnet rotates around a coil, producing a current that may be measured. If a tachometer is not attached to the centrifuge, a strobe light or an electronic meter can be used to determine the actual rpm of each centrifuge setting. For any given setting, the speed of the centrifuge will vary with specimen load. The rpm determined for each centrifuge setting should be recorded and attached to the centrifuge for easy reference. However, the value of rpm is only one factor involved in determining centrifugal force. The true efficiency of a particular centrifuge must be expressed in relative centrifugal force, some number x gravity. The relative centrifugal force is calculated from the experimentally determined rpm and the radius of the head plus carrier. The RCF value for various speed settings is also generally listed in tabular or graphic form in the manufacturer's instruction manual for the instrument. The term rpm, when applied to a centrifuge, does not give any indication of the force applied to samples in that centrifuge. All comparisons made between centrifuges should be in relative centrifugal force.

A timer device is commonly employed in many centrifuges. Most are spring-driven clock mechanisms which turn the unit off after a present time cycle. Some units have an electronic timer which performs the same function. Many units have a switch that allows the unit to be operated continuously without using the timer device.

Braking devices are incorporated to provide rapid de-acceleration of the unit. Two general types of brake mechanisms are found on centrifuges. One type is mechanical and functions by physically applying pressure to the rotor at some point when a lever is pressed. The other type, found on most units, is electrical in nature and functions by reversing the polarity of the current to the motor.

Rotors

Centrifuge rotors are usually designed either for preparative or analytical purposes. Most analytical rotors are constructed so that light, which is used to determine the position of the solute boundary, can pass through the sample and its container while it is spinning in the rotor. Details of analytical rotors are beyond the scope of the present text but may be found in most of the references to ultracentrifugation.

Preparative rotors are of three principal types: (a) angle, (b) swinging-bucket, and (c) zonal. Angle and swinging-bucket rotors are much more common than the zonal type. In the angle rotor, containers, which may be either tubes or bottles of glass, metal, or plastic, are inserted into cavities in the rotor that are constructed at an angle to the axis of rotation. In the case of the swinging-bucket rotor, the container is placed in a holder which is itself held to the rotor body by means of pins that allow it to swing. The vertical axis of the container is parallel to the axis of rotation when the rotor is at rest, but as it begins to spin the container and its holder swing out because of the centrifugal force. At operational speeds the container is usually perpendicular to the axis of rotation.

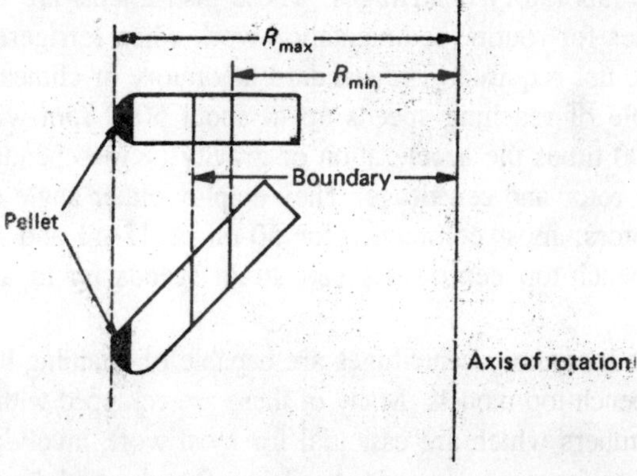

Fig. 3.4. Relationship of R_{max} and R_{min} to the axis of rotation in swinging-bucket and angle centrifuge rotors.

In addition to the three types of rotors mentioned above, a special type of rotor is used in preparative centrifuges manufactured by the Sharples Equipment Co. This type of rotor, which is a hollow stainless-steel cylinder, has a bottom fitted with crossed baffles. As the rotor spins in the centrifuge, the solution containing the particles flows from the bottom across the baffles into the interior of the rotor. The vortex of liquid created by the spinning baffles causes the heaviest particles to sediment onto the lower wall of the rotor whereas the lighter particles sediment higher on the wall. If the speed and flow rate are controlled properly, the solvent and the lightest particles flow out through holes in the top of the rotor. This type of rotor is very effective for *continuous-flow centrifugation* and can be used to harvest large batches of microorganisms from their liquid cultures and subcellular particles from large batches of solution. Angle centrifuge rotors have also been adapted for continuous-flow centrifugation, and several models are available commercially.

All rotors must be designed so that they do not come apart under the stress of centrifugal fields. They should be constructed of a strong metal and should be machined so that their opposite sides strike a balance in weight. Those rotors which are used for very high speeds must be stronger and better-balanced than those used at low speeds. In addition, the metal used for the rotor should be protected with a resistant coating.

Types of Centrifuges

Most students of biochemistry are already familiar with standard bench-top laboratory centrifuges. These instruments are used in most laboratories for routine sedimentation work when refrigeration or high speeds are not required. The standard laboratory or clinical centrifuges are capable of reaching speeds up to about 5000 rpm with forces of about 3500 times the acceleration of gravity (×g) depending upon the particular rotor and centrifuge. They employ either angle or swinging-bucket rotors, most commonly for 50-ml or 12-ml and 15-ml tubes. Certain bench-top centrifuges can attain speeds up to about 16,000 rpm.

Other laboratory centrifuges are capable of attaining higher speeds than the bench-top models. Many of these are equipped with refrigerated rotor chambers which are essential for most work involving enzymes. These centrifuges are usually available for use with many different rotors that offers a wide choice of speeds and sample volumes.

The most highly refined centrifuges are called ultracentrifuges. These instruments are available in both preparative and analytical models although, with the proper techniques, preparative models can be used for analytical purposes. Ultracentrifuges are capable of attaining very high speeds, up to about 75,000 rpm, with centrifugal forces of up to about 500,000 x g. Since such speeds result in consideration heating of the rotor, because of the friction with air, ultracentrifuges are usually constructed with a vacuum rotor chamber and refrigeration. Rotors for use at the highest speeds are constructed of titanium.

Analytical models of the ultracentrifuge are equipped with optical systems to measure the position of the solute boundary during centrifugation. These systems utilize refractive or absorptive properties of the sample to make measurements. Many analytical ultracentrifuges are also equipped with controls which enable them to be operated at carefully regulated slow speeds so that solutes may be studied by the approach-to-equilibrium technique. These systems and methods of analytical ultracentrifugation are discussed extensively in the references to ultracentrifugation found at the end of this chapter.

Several different centrifuge models are designed for or are modified for continuous-flow sedimentation. These either utilize angle rotors or the Sharples cylindrical type of rotor discussed earlier.

Most centrifuges are equipped with electrical motors to propel the rotors. A few centrifuges utilize compressed air or steam for this purpose.

General Operation and Maintenance

Daily inspection along with periodic function verification and proper preventive maintenance are vital to the efficiency and longevity of the centrifuge. A regular schedule for checks must be established and followed to insure proper operation of the instrument.

Daily Operation

Daily operation should include observation and inspection of the following:

1. The centrifuge should not be on the same circuit as sensitive electronic measuring devices such as spectro-photometers, since it generates electrical noise and has a high current drain at start-up.
2. Check the cleanliness of the chamber, and immediately clean up all spills. Be aware of biohazardous materials (microbiologic, radioactive, chemical) which require specific decontamination procedures.
3. Always balance the load of the centrifuge before operating: use the correct tube sizes and type for the particular centrifuge.
4. Always insure that the cover is closed and latched while the unit is operating. This will prevent the dangerous scattering of both biohazardous and physically dangerous material out into the environment where the operator and others may be exposed.
5. Observe for unusual noises or vibrations during operation.

Function Verification

Frequency of function verification procedures should be appropriate for the application of the centrifuge. It is generally recommended that procedures be performed at 3-month intervals with more frequent checks on those units with critical application. All checks performed and data gathered should be recorded in such a fashion that the information is readily available to the operator. Correction factors for speed, timing, or temperature should be posted on the unit.

Tolerance limits for function checks should be established. The limits set will again depend on the application of the centrifuge. Include information for corrective action should tolerance limits be exceeded. Function verification procedures should include the following:

1. *rpm calibration*. One of the best ways to check the function of a centrifuge is to check its speed. Depending on how the unit is equipped, both the speed control and built-in tachometer should be checked with an external device. This may be accomplished with either a strobe light or mechanical or electronic tachometer of good accuracy. Several speeds used regularly on the unit should be checked. Values obtained with the external measuring device should agree within 5% of those with a built-in tachometer.
2. *Timer*. The timer should be set for common timing intervals and these checked against an accurate stop-watch or electronic timer. General laboratory centrifuges should be accurate to 10% of the total timed interval.
3. *Temperature*. The thermometer on refrigerated units should be checked against a certified thermometer, and a correction factor should be derived if necessary.

Preventive Maintenance

Preventive maintenance should be performed on the same time schedule as are function verification procedures. The preventive maintenance schedule should include the following:

Lubrication

Depending on the type of centrifuge, bearings on the upper and lower end of the motor shaft may be permanently lubricated or sealed. If this is not the case, then the manufacturer's instructions must be followed for lubrication. Bearing wear may be checked at this time by determining the amount of side play in the shaft.

Motor components

Brushes should be removed and checked for wear. Replacement is recommended if they are worn to more than one-half their original length. When reinserting used brushes, replace them in the same orientation, and be certain that spring tension is adequate to maintain good contact with the commutator. The condition of the commutator should be examined. In order to avoid electrical arcing, the commutator and brush holders must be free of dirt, oil, and dust. If the commutator is scartched or scored, it will have to be removed and machined smooth with a lathe. On refrigerated units the manufacturer's instruction manual should be consulted for maintenance procedures on the refrigeration system.

Electrical integrity

Both grounding resistance and current leakage should be checked periodically for proper rating; if the unit has a circuit breaker, it should be checked for proper operation. In addition, the line cord, plug, lamps, and wiring should be examined for defects.

Mechanical integrity

If the unit is equipped with a safety interlock, this device should be checked for proper working order. Gaskets, latches, hinges, and control knobs should be examined to determine that all are functioning and in good condition. The head and shields or carriers should be examined for signs of mechanical stress (cracks) and for cleanliness and balance.

Use and Care of Rotors and Centrifuges

Before using any centrifuge read the operational directions or obtain instructions form an experienced operator. Observe the following rules:

1. Do not overfill centrifuge tubes or bottles.
2. Balance the containers with their solutions before placing them in opposite cavities of a centrifuge.
3. Check the rotor to see that it is empty and undamaged before beginning a centrifuge run.
4. Do not exceed allowable speeds for a given rotor.
5. Turn off the centrifuge power immediately if any irregular noise or vibration occurs.
6. Keep rotors and rotor chambers clean. Chambers can be cleaned with a sponge, warm water, and a mild detergent; rotors can be cleaned with a bristle brush, warm water, and a mild detergent followed by rinsing with distilled water and drying by resting them upside-down.
7. See that the centrifuge is maintained according to instructions.

Some rotors for ultracentrifuges suffer from metal fatigue with prolonged use. Therefore, after a certain amount of use their maximum allowable speed must be decreased, a procedure called derating. For this reason, a record of their use must be maintained. Instructions on derating are provided with these rotors.

4

PRIMARY CELL LINE

Primary cell cultures are those that are composed of cells taken directly from a living animal. Primary cells normally contain a diploid set of chromosomes, have limited lifespan and undergo aging. In order to prepare a primary culture, an organ or tissue is removed from a freshly sacrificed animal and aseptically cut into small pieces and treated with enzymes such as trypsin, collagenase, pronase or combination thereof. After the cells have been separated from the tissue, they are inoculated into appropriate cell culture medium in tissue culture flasks. The cells will adhere to the surface of the flask and replicate until they come in contact with each other. Thus they attach and grow as a uniform layer of cells, or a monolayer, which is always one cell thick. These cells will stop growing once the surface is filled up and contact follows. This is known as contact inhibition. Cells can be cultured as stationary monolayers usually inoculated with 2 x 10^5 cells in 5 ml of cell culture medium in a petri dish designed for tissue culture work.

Some cell types can be grown in suspended state in culture medium. Suspended cells are generally derived from blood cells. B lymphocytes can be established in culture directly from leukemic cells or by infecting normal lymphocytes with Epstein-Barr virus. Suspension cell cultures will yield 5-10 times more cells per ml of medium than monolayer cultures.

It is fairly easy to derive a primary culture of fast growing cells like fibroblasts, and as a result most primary cultures consist of fibroblasts. Fibroblasts are connective tissue cells which secrete the extracellular matrix of connective tissue. In this exercise, primary fibroblast cultures will be derived from chicken embryo. They will be made by dissociating the entire embryo with the proteolytic enzyme trypsin. Embryo cells are somewhat easier to adapt to cell culture than adult cells, since embryo cells are less differentiated and more likely to be rapidly dividing. The culture that is first produced will contain a variety of cell types. It is possible, however, to eventually establish a homogeneous culture of fibroblasts, because the other cell types do not grow as fast and they will be diluted out in subsequent passages subcultures.

Safety Guidelines

Wear gloves, lab coat, and safety glasses. Standard laboratory safety procedures should be followed.

Experimental Outline

- Establish primary cell culture from chicken embryo.
- Culture cells as stationary monolayer in tissue culture petri dishes.
- Establish a homogeneous culture of fibroblasts.
- Grow cells in tissue culture medium under sterile conditions.
- Stain cells with trypan blue and count the number of unstained, live cells under a microscope using a hemocytometer.

Materials

Fertile chicken eggs, incubated 7-10 days.

A bottle of 0.25% trypsin solution.

A bottle of PBS (phosphate buffered saline).

Tissue culture flask 25 cm^2.

A bottle of HMEM (Hank's minimum essential medium) + supplemental 10% Fetal Bovine Serum (heat inactivated).

Three 100 mm plastic petri dishes.

One 50 ml plastic beaker.

150 ml squeeze bottle with 70% alcohol.

A sterile dissection kit containing sharp scissors, sharp scalpel and two pointed forceps.

95% ethanol.

5 ml plastic syringe.

Sterile gauze.

Two 125 ml flasks.

Trypsinizing flask.

15 ml centrifuge tubes (2).

Trypan blue (0.4%).

0.15 ml plastic microcentrifuge tube with an attached cap penicillin-streptomycin 100x stock solution.

1 ml and 5 ml pipets, sterile.

Pipet pumps.

CO_2 incubator.

Method

1. Keep 0.25% trypsin, HMEM and PBS in 37°C water bath.
2. Aseptically place 15 ml of PBS into each sterile petri dish.
3. Place the egg in a 50 ml beaker with its blunt end up, and disinfect the entire egg shell with 70% ethanol.
4. With sharp sterile forceps puncture the top of the shell and remove the shell all the way up to the base of the air cell.
5. Locate the position of the embryo and carefully tear the membrane away from the embryo with sterile forceps.

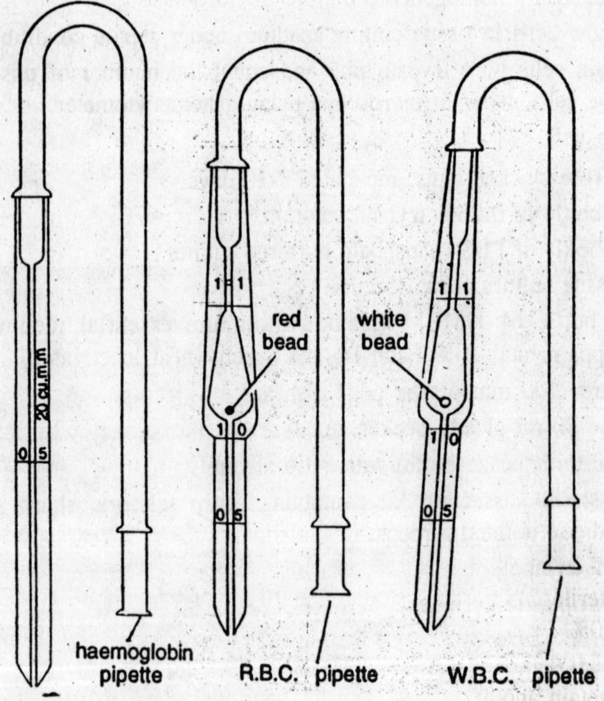

Fig. 4.1. R.B.C., W.B.C. and haemoglobin pipettes.

6. Insert a sterile pair of forceps into egg and grab embryo firmly. Take it out of egg and place in the first dish of PBS.
7. Cut off the head and feet of embryo with a sterile scissors. This procedure is done for 10 day old embryos but not for younger embryo.
8. With a flamed sterile forceps, transfer the remaining parts of the embryo to a petri dish containing PBS and rinse thoroughly. Repeat the rinse until no color is found in the rinse solution in the dish.
9. Take a sterile 5 ml syringe and a sterile capped test tube. Remove the plunger from the syringe and rest it on sterile kimwipes. Use a sterile forceps to transfer the remaining parts of the embryo from the petri dish into the syringe. Remove the cap from the tip of the syringe without touching the tip with your fingers. Uncap the sterile test tube and place the cap on sterile kimwipe and keep the tube in a test tube rack. Position the tip of the syringe over the test tube. After inserting the plunger back in the syringe, with a steady and even pressure, push the plunger all the way down so that the embryo is forced through the syringe into the test tube. The embryonic tissues are delicate and are homogenized to very fine pieces.
10. Transfer the tissue into a trypsinizing flask and add 10 ml of 0.25% prewarmed trypsin solution.
11. Place the trypsinizing flask on the shaker in the 37°C water bath and allow the suspension to swirl on the shaker at low speed so that the tissues of the embryo just barely graze the cutting edges of the trypsinizing flask. It will be necessary to continue this action for 15-60 minutes depending on the disaggregation process, and you will have to monitor this process closely. The flask is removed once the trypsin solution becomes cloudy with individual cells. Excessive trypsinization will reduce the cell viability significantly.
12. Allow fragments to settle; the supernatant is a cloudy suspension of cells. Pour this cell suspension through the side spout into a sterile 125 ml flask containing 10 ml of prewarmed HMEM with 10% Fetal Bovine Serum. The serum will inhibit further action of trypsin and prevent further damage to cells.
13. Obtain another sterile 125 ml flask and place two or thee layers of sterile gauze over the mouth of the flask and filter the cell suspension through this gauze to remove large cell clumps and other debris.

Fig. 4.2. The Burker Hawksley counting slide.

14. Take two clean sterile centrifuge tubes. Aseptically transfer the cell suspension into these tubes so that both tubes contain the same quantity. Pellet the cells by centrifugation in a table top centrifuge for 5 minutes at 1000 rpm. Discard the supernatant and resuspend the cell pellets in 5 ml of HMEM containing 10% Fetal Bovine Serum. Combine the cell suspension in one tube and cap it.
15. Remove the cap of the tube containing the cell suspension and by using a sterile 10 ml pipette, aspirate the suspension by drawing it up into the pipette and then forcing back into the tube. Do this about six times. It is important to keep the cell suspension uniform and the clumps broken up as much as possible. Determine the viability and cell density.
16. With a sterile 1 ml pipette, aseptically transfer 0.1 ml of cell suspension into 1.5 ml microcentrifuge tube.
17. Pipette out 0.9 ml of 0.4% (w/v) trypan blue solution into the above microcentrifuge tube (wear gloves when working with trypan blue). Cap the tube tightly and vortex it for 15 seconds. Immediately with a Pasteur pipette introduce a drop of it at the center of one edge of the cover glass covering one of the counting chambers of the hemocytometer. The fluid will be drawn under the cover glass by capillary action. Do not over fill the chamber.
18. Place the counting chamber on a regular microscope and focus with low power objective on the ruled area of the chamber. The hemocytometer is divided into 9 large blocks (1 mm^2), and each of these large blocks is divided into 16 smaller squares. Count the number of unstained cells (trypan blue penetrates damaged cells, and thus only stains those cells that are dead) in the central large block and four large corner blocks. If there are too many cells to count, make the necessary dilution.

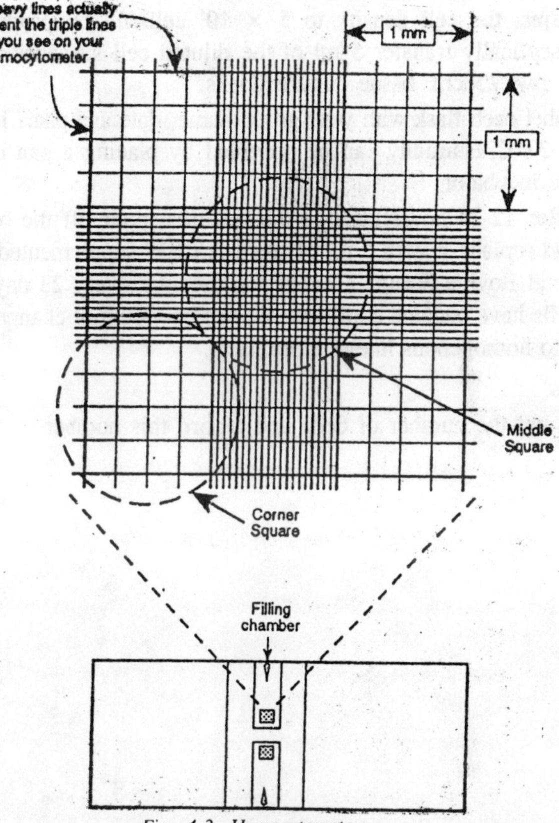

Fig. 4.3. Hemocytometer.

19. Divide total number of cells by 5 to obtain average number of cells per large block. (Note: if this number is smaller than 10 and larger than 100, the count will not be accurate.) Multiply by 10^4 and the dilution factor to obtain number of cells per ml in the suspension. An example is shown below:

    Total cells counted in give large blocks = 250
    Average number of cells/block = 50
    Number of cells per ml = 50×10^4

    For the correct count, multiply the above number by 10 (dilution factor). The correct number of cells would be:

    $$50 \times 10^5$$
    $$5 \times 10^6$$

20. Adjust the cell density to 5×10^5 cells/ml by proper dilution. Aseptically transfer 5 ml of the diluted cell suspension into each of two 25 cm^2 tissue culture flasks.
21. Label each flask with your group name, date and place in incubator at 37°C Humidity can be provided by placing a pan of water in the incubator.
22. After 12-24 hours observe the cultures, pour off the old medium and replace it with 5 ml of fresh HMEM supplemented with 10% Fetal Bovine Serum. Change the medium every 23 days until the cells have become confluent. Notice how culture changes overtime into homogenous line of fibroblasts.

Results

Count the number of cells and record this number.

5

PLANT CELL CULTURE

When a plant is wounded, cells at the wound site proliferate into an undifferentiated mass called *callus tissue*. This tissue can be thought of as a kind of scar tissue and, in nature, callus formation seals the wound against attack by pathogenic bacteria and fungi. Callus is probably most familiar to you as the swollen, knot-like growth seen on trees at the site of limb removal, but herbaceous plants as well as woody plants produce callus. Plant cell culture exploits this natural wound response of plants by encouraging callus growth on small pieces of excised tissue or "explants".

On artificial media, callus can be maintained in a state of persistent, undifferentiated growth. By changing the concentration of plant hormones in the medium, however, the callus of many plant species can be made to redifferentiate into whole plants. The ability of single cells to regenerate whole plants is referred to as *totipotency*.

The commercial applications of plant tissue culture techniques are quite important. Regeneration of plants from cell culture offers a practical strategy for plant cloning, since all regenerants from a culture should be genetically identical. For commercially valuable plants that are difficult, costly or inefficient to propagate by cuttings or other asexual means, cell culture sometimes offers the only practical means of propagation. Most hybrid orchids, for example, are propagated today by the tissue culture method of "meristemming" or "mericloning". Plant cultures are also being investigated as sources of valuable plant products like drugs, flavors and fragrances.

Many strategies for genetic engineering of plants rely on plant tissue culture. Plant cells in culture can be genetically transformed by a number of techniques. Explants can be transformed directly by *Agrobacterium tumefaciens* or by bombardment with DNA-coated particles from a particle gun. Protoplasts (plant cells with their cell walls removed) are the targets of microinjection and electroporation (DNA uptake mediated by an electric field). Plant cell culture is central to these techniques in that cell culture allows transformants to proliferate and, sometimes, regenerate into genetically identical clones.

While many formulations for plant tissue culture media have been developed, all contain the same basic types of ingredients. These are:

1. Inorganic salts
   Macro and micronutrients
2. Vitamins
3. An organic carbon source

In addition to these components, most formulations contain plant growth regulators (plant hormones), *auxin* and *cytokinin*. Callus media are generally solidified by the addition of agar. (Plant cultures are sometimes maintained as a suspension of cells in liquid medium, but this type of culture will not be included in this laboratory experiment.)

The medium used in this experiment, MS medium, is a formulation developed by Murashige and Skoog (1962.) MS medium is one of the most commonly used of all plant culture media. The ingredients included in this medium are:

1. Inorganic Salts	mg/l
Macronutrients	
Ammonium nitrate	1650
Potassium nitrate	1900
Calcium chloride (anhydrous)	332
Magnesium sulfate (anhydrous)	180.7
Potassium phosphate	170
Micronutrients	
Ethylenediaminetetraacetic acid (EDTA, Disodium Salt)	37.3
Ferrous Sulfate (heptahydrate)	27.8
Manganese sulfate	16.9
Zinc sulfate (heptahydrate)	8.6
Boric acid	6.2
Potassium iodide	0.83

Sodium molybdate (dihydrate)	0.25
Cobalt chloride (hexahydrate)	0.025
Cupric sulfate (pentahydrate)	0.025
2. Vitamins	
Myo-inositol	100.00
Thiamine hydrochloride	0.40
3. Carbon source	
Sucrose	30,000.00

The medium is solidified with 7.5 g/l of agar. The amounts of plant growth regulators added to the medium are variable, depending on whether the culture is to be maintained as callus or made to regenerate whole plants. The medium used in this experiment contains Indole Acetic Acid (IAA, an auxin) at a concentration of 1.0 mg/l. No cytokinins are added. The effects of varying the concentrations of plant growth regulators on the cell culture will be investigated in another module.

Safety Guidelines

Follow standard laboratory safety practices.

1. Preparation of media requires an autoclave or pressure cooker. Use of this equipment requires care to avoid serious burns.
2. An alcohol or gas flame is used to sterilize instruments during the lab exercise. Due care should be used with open flames. Precautions should be taken to avoid igniting hair or clothing.

Experimental Outline

1. Disinfect plant tissue and prepare explants.
2. Inoculate medium with plant tissue.
3. Observe cultures over a period of several weeks.

Materials

Sterile media (this has been prepared prior to class)
Plant material
10% Household bleach solution (prepare fresh)
Sterile Water
70% Ethanol
95% Ethanol
Forceps
Scalpel

Pre-lab Preparation

Tissue culture medium should be prepared in advance of the laboratory period. This can be done several days to a week ahead of time. Prepared media can be stored at room temperature. Plant materials used for the experiment should be purchased or collected shortly before the class meeting.

Preparation of plant material for culture, i.e. surface sterilization, will take approximately 30 minutes. Plan another 30 minutes for novices to prepare explants and inoculate the culture medium. The time required for students to complete the exercise will depend on whether students are waiting for space in a laminar flow hood.

Depending on the length of the laboratory period, you may want to combine this exercise with Module 3: Regeneration and Cloning. Both exercises require that the students initiate cell cultures.

Because results from both Modules 1 and 3 require several weeks of observation after the initial laboratory class period, these exercises are best used early in the semester.

Timetable of events

Students will learn how to establish plant cell cultures from a variety of plant tissues. In accomplishing this, students learn how to prepare plant tissue for culture, practice sterile technique, and learn about the components of plant culture media.

The "bare bones" methods described in the student guide for this module (and for others in the plant tissue culture unit) are adaptable to most lab conditions. Depending on the availability of sterile transfer hoods, specialty glassware, etc., you may want to elaborate on the protocols presented, but very little specialized equipment is actually required for success.

Equipping the work area

Plant tissue culture is normally done in a laminar flow hood, a containment hood, or in a transfer cabinet. The laminar flow hood has a box shape with one open side facing the worker. Fans in the hood circulate air through a particle (HEPA) filter which removes airborne contaminants. Air flow in the laminar flow hood is from the back of the box outward past the worker. Sterility in the work area is maintained by this flow of clean air, but can be compromised by introducing contaminated equipment into the hood, by leaning into the hood while working, or by placing non-sterile items between open cultures and the back of the hood.

Containment hoods, like laminar flow hoods, depend on air flow to maintain sterility in the work area. Because these hoods are designed to prevent the escape of cultured cells from the work area, however, the opening at the front of the hood is partially closed (generally by a window) and air travels from the top of the cabinet downward to intakes on the work surface. Air does not flow out of the hood into the lab environment without first being filtered to remove contaminants. The protection offered to workers by a containment hood is unnecessary for most plant tissue culture purposes, but the hood is very satisfactory for handling plant cultures.

Transfer cabinets (or closely-related glove boxes) are box-like enclosures, sealed on the top and on three sides, with limited access from the front. Sterility is maintained within the cabinet only to the degree that air movement from outside the box to the inside is restricted. The transfer cabinet is really no more than a shelter against airborne contaminants. When used properly, the transfer cabinet is an economical and effective enclosure for cell culture. If desired, a transfer cabinet can be economically "homemade" from plywood or plastic. Painting the plywood surface with enamel will make it washable.

In the absence of a sterile hood or transfer enclosure, plant culture experiments can still be carried out. However, depending on the cleanliness of the lab environment and the sterile technique of the students, losses to fungi and bacteria can be expected. The best results will be obtained if windows are kept closed to restrict air movement and if bench tops are wiped with 70% ethanol before work begins. Perhaps most important of all, students should wash their hands well before doing any culture work.

The equipment required for plant cell culture is modest. For handling cultures, the work station should be outfitted with a Bunsen burner, a covered container of 95% ethanol, forceps (long enough to reach into culture flasks or tubes), and a sharp scalpel. For disinfecting the work area, a wash bottle or spray bottle of 70% ethanol and a supply of clean (preferably autoclaved) wipers or paper towels will suffice. For media preparation, you will need a balance weighing to 0.1 mg. a pH meter, and an autoclave. Plant cell cultures can be grown in petri dishes, baby food jars, test tubes, Erlenmeyer flasks, mason jars, etc. The only requirements for a suitable container are sterility (they should be autoclavable or presterilized/disposable) and they must be able to be opened and closed repeatedly without introducing contaminants. Cotton or foam stoppers can be used as closures. Specially-designed caps for tubes, flasks and baby food jars are also available. Ideally, plant callus cultures are incubated at 26-28°C, but they will grow at slightly cooler room temperature. Callus cultures can be maintained either in the light or in the dark. In the classroom situation, a drawer in the lab bench can serve the purpose of incubator. Callus cultures from which plants are being regenerated should be maintained under fluorescent lights with a day length of 16 hours.

Preparation

Obtain plant material: A number of different plants can be tested in this experiment. Students may even want to try some materials of their own. The best material for explants comes from plants raised indoors or in the greenhouse because they seem to be "cleaner". Some vegetable tissues are notable exceptions to this. Some plants to try are: Tobacco stem or leaf, carrot roots, African violet leaf or petiole, cauliflower florets. If you want to use vegetables from the grocery, try to purchase these as fresh as possible. Plants other than those listed above will also do very well. Make the selection of explants part of the experiment.

Make media: The recipe for MS medium is written out in the student guide to the laboratory. If you choose to assemble the medium from stock bottles on the lab shelf, it will be easier to prepare a 100 x concentrated stock solution of ingredients listed as micronutrients and keep this stock in the refrigerator. Also, plant growth regulators are more easily handled as stocks. Dissolve IAA in a small amount (few mls) of 1 M NaOH, then dilute to 1.0 mg/ml with distilled water. Dissolve kinetin in a small volume of 1 M HCl, then dilute to 1.0 mg/ml with distilled water. Just before autoclaving, adjust the pH of the plant medium to 5.8.

Pre-prepared media are available commercially and offer a simple and cost-effective alternative for media preparation. For this experiment, order "MS salts with Minimal Organics". To this powder, you will only have to add sucrose (table sugar from the grocery store will work; add 30 g/l), agar (7.5 g/l); and plant growth regulators (For this experiment, add only IAA; 1 mg/l).

Method

1. Prepare plant material for culture. Select fresh-looking, healthy (not brown, bruised or wilted) plant material. This can be leaf, stem, or root. Cut the tissue into manageable pieces. These should

be small enough to fit into a beaker for disinfecting, but large enough so that they can be trimmed after disinfection.

2. Move your work into a sterile transfer hood. If a hood is not available and you are working at the lab bench, wipe the work area with a solution of 70% ethanol before beginning and try to keep your work covered as much as possible. Before proceeding to disinfect your plant tissues, wash your hands thoroughly.
3. Wash the explants in a beaker of distilled water to which you have added a few drops of detergent.
4. Using forceps, transfer the explants to a 70% solution of ethanol for 2 minutes.
5. Again using forceps, transfer the explants to a 10% solution of household bleach for 5-10 minutes. Tender leaf tissue should not be left in the bleach for more than about 5 minutes. Root or stem pieces can be left for longer times.
6. After treatment with bleach, the tissue is considered to be sterile. Care should be taken to avoid re-contaminating it. At your work station should be a covered container of 95% ethanol and a Bunsen burner. Before they are used to handle sterile plant tissues, forceps and scalpels should be dipped in ethanol and passed through the flame several times to sterilize them. Be careful as you do this. An alcohol flame is nearly colorless and therefore invisible. Take precautions to avoid igniting hair or clothing. If you should accidentally set the container of ethanol on fire with the hot instruments, extinguish the flame by replacing the cover on the container.

   Note: Instruments are sterilized by ethanol and not by heat or flame.
7. Rinse the explants in three, 5 minute changes of sterile distilled water. Keep the tissue in its last rinse until you are ready to use it.
8. Transfer the explant from the rinse water to the lid of a sterile petri dish. For leaf tissue, cut the tissue into small squares, not larger than 1 cm^2 using a flamed scalpel. Cut edges that were in direct contact with ethanol and bleach should be trimmed away. Do not use leaf tissue that appears "soaked through" by the disinfectant. Root or stem tissue should be trimmed to remove any tissue that was in direct contact with the disinfecting solutions. Cut the tissue into small cubes, 0.5-1.0 cm^3. In preparing your

explants for culture, be mindful of the need to maintain their sterility. Flame your forceps and scalpel frequently and don't lay them down on the work surface after flaming. When working in a laminar flow hood, remember that it is important not to place any contaminated materials up wind of your work. If you are not working in a hood, work quickly and keep your work covered as much as possible.

9. Using flamed forceps, transfer the explants to the culture medium. Be careful to prevent contaminating the medium during this procedure. Explants should be placed firmly in contact with the medium, but should not be buried in it.
10. Label your cultures as to source of explant (type of plant and tissue), date, and your name. Consult the instructor about where cultures are to be incubated.
11. Check the cultures periodically before your next lab class. Any cultures that become grossly contaminated should be autoclaved before disposal.

Results

Examine your cultures and others. Use a microscope. Describe the callus tissue formed and determine the following. How do callus cells compare in appearance to the cells from which they grew? Is there any difference in the appearance of the cultures developed from different plant species or from different tissues of the same plant? Did all of the cultures grow equally well? Did callus tissue form at the same time and grow at the same rate in each of the cultures? Did any of the cultures begin to regenerate roots or shoots? What types of contaminants appeared in the cultures?

Within the first week explants can be expected to swell noticeably. Callus formation should begin during this time. Callus should appear on cut edges of the explant and will look like a rough white fringe of cells. In subsequent weeks, the callus will proliferate into an undifferentiated mass of cells. If desired, the callus can be cultured away from the explant. Remove the clump from the explant tissue with sterile forceps and place it on fresh medium. Whether or not the callus is to be maintained, students should explore the culture with forceps to get a feel for the texture of this tissue. Depending on what plant material is used for the cultures, shoots or roots may begin to regenerate from the callus. This is not a likely outcome, but it is possible.

The major difficulty faced in initiating plant tissue cultures is that of contamination. Plant media are excellent substrates for a host of fungi and bacteria. Generally speaking, bacterial contaminants are more easily dealt with. Antibiotics such as cefotaxime (100 μg/ml) or piperacillin (100 μg/ml) can be included in media for short term bacterial control. If you choose to use antibiotics, remember that they should not be added to media before autoclaving. They should be added as a filter-sterilized stock solution to cooled medium immediately before it is dispensed into tissue culture vessels. Fungal contaminants are nearly impossible to get rid of. For the teaching lab, probably the best advice about contamination is to tolerate it if possible (that is, if the cultures aren't grossly contaminated) over the short term of the experiment and dispose of contaminated cultures by autoclaving. Sometimes, contaminants will not interfere with the outcome of the lab exercise.

6

FORMATION OF PROTOPLASTS

Protoplasts are plant cells whose cell walls have been removed. Recall that plant cells *in vivo* are surrounded by rigid cell walls composed largely of celluloses and cemented together by pectins. The cell wall normally confines the cell, preventing it from bursting as a result of turgor pressure. For some biotechnology and genetic engineering protocols, however, the cell wall is a substantial barrier. Releasing plant cells from the confines of the cell wall allows them to be manipulated by microinjection, electroporation and fusion.

Many techniques for the production of protoplasts have been developed. Originally, tissues were bathed in a solution of high osmotic potential to shrink the cells. Then, cell walls were broken mechanically by cutting or abrasion. Finally, by reducing the osmotic potential of the medium, the cells were made to swell and pop out of the broken cell walls. The yield of viable protoplasts from this type of protocol is generally low, thus, except for some difficult tissues, this strategy is rarely followed. Most methods today depend on the enzymatic breakdown of cell walls first demonstrated by Cocking (1960). Protoplast digestion mixtures generally include several enzymes, the concerted action of which efficiently releases cells from the cell wall. A mixture of cellulase, pectolyase, and macerase will be used in this experiment. Because naked plant cells are fragile, and can burst in solutions of low osmotic potential, mannitol is included in the digestion mixture as an osmoticum. Other ingredients included improve the stability of the protoplasts.

The plant material used in the experiment is red onion. From this plant, you will prepare protoplasts from the bulb (a modified leaf structure). Because not all cells of red onion are pigmented, protoplasts are a mixed population of colored and colorless. After preparing the protoplasts, you will use them to perform a cell fusion experiment. By treating the protoplasts with solutions of polyethylene glycol and calcium, they can be made to fuse to form hybrid cells. These cells will be recognizable under the microscope. Can you think of any possible applications of this procedure in biotechnology?

If your purpose in this experiment were to grow up the hybrid cells or even to regenerate whole plants from them, it would be necessary for you to isolate the protoplasts from sterile (i.e. axenic) tissue, to maintain sterility throughout the protocol, and to return the hybrid cells to culture where they could continue their development. These manipulations are not trivial. In practice, a protoplast fusion experiment of the type you are doing here would be done on a much larger scale and the protoplasts would be treated somewhat differently. First, the protoplasts to be fused would likely originate from different tissues, generally from different genotypes or even different species. The protoplasts would be washed after isolation to remove traces of the digestive enzymes used to remove their cell walls. After fusion, they would be plated in media by techniques that encourage them to regenerate cell walls. In this laboratory experiment, the precautions necessary to maintain your protoplasts in long term culture will not be followed, but you should understand how this experiment would be different if it had different goals.

Safety Guidelines

Follow standard laboratory safety practices.

Experimental Outline

Cut up onion tissue

↓

Incubate in digestion mixture (1 hr)

↓

Collect protoplasts, observe, allow to settle on slide (20 min)

↓

Add fusion buffer and incubate (20 min)

↓

Add W10 buffer and observe fusion (20 min)

Materials

Plant material: Red onion. Choose a firm unblemished bulb.

Petri dishes or depression slides: Small diameter (60 mm) disposable dishes or microscope slides with depressions or wells are the container of choice for digesting the plant tissue. Small test tubes will work, but petri dishes offer the advantage that they can be viewed on a microscope so that students can monitor protoplast release.

Microscope: An inverted scope, of the type normally found in tissue culture labs, is the instrument of choice for this use. If such a scope is not available, the normal student scope will work if the experiment is done on a microscope slide. Depending on the magnification available, even a dissecting scope can be used. Since the success of this experiment depends on the students' viewing the results, test available equipment before attempting the lab exercise.

Pasteur pipets

Enzyme solution:

1.5% Cellulysin
1.5% Macerase
0.2% Pectinase
0.4 M Mannitol
2.0% Glycine

PEG Fusion Buffer:

0.3 M glucose
66 mM $CaCl_2$
40% polyethylene glycol (molecular weight 1400)-pH 6.0

Solution W10:

9 parts solution A: 0.4 M glucose
66 mM $CaCl_2$
10% dimethylsulfoxide

1 part solution B: 0.3 M glycine-OH buffer pH 10.5

Pre-lab Preparation

Timetable of events

Prepare protoplasts (1 hr)

Do fusion experiment (45 min)

Filter the enzyme solution through a 0.45μm filter (Gelman) before use. The solution can be stored frozen for several months without significant loss of activity. Also prepare some of the same solution without enzymes for use as "medium" in step 6 of the student's experimental procedures.

Method

1. Using a razor blade or scalpel, cut a small square of tissue (about 1 cm^2) from one of the layers of a red onion. You will notice on the cut edge that the tissue is only red on one surface. Slice the tissue again, layer cake fashion, to expose cells that were buried in the center of the square.
2. Place the half of your tissue containing both red and white cells face down in a small pool (about 0.2-0.5 ml) of enzyme solution. Cover the container in which the tissue is being digested to prevent the enzyme solution from evaporating. Allow the enzyme to work for 1/2 to 1 hour. You will want to examine the digestion periodically during this time to monitor the release of protoplasts. It will help to occasionally agitate the mixture gently.
3. At the end of 1 hour, remove with forceps any remaining bits of undigested tissue. Remove 2 drops of protoplasts to a clean microscope slide. Protoplasts are spherical structures. Do not cover them with a coverslip. Check the slide on the microscope to see that you have both red and colorless protoplasts. The actual number of protoplasts on the slide is not critical, but there should be enough in your sample that they are not difficult to find. Typically, several dozen should be readily visible. If this is not the case, make a new slide. (The protoplasts are actually large enough so that you can see to collect them with a pipet.)
4. Place the slide on the lab bench where it will not be disturbed and allow the protoplasts to settle onto the glass about 15-20 minutes.
5. Add 1 drop of PEG Fusion Buffer to the protoplasts. Check the slide on the microscope at this point. You should see the protoplasts clumping together. Wait 15 minutes, then carefully remove the liquid from the slide with a pipet. The protoplasts should remain behind.
6. Add 2 drops of solution W10 to the slide. Check the slide on the microscope periodically over the next 20 minutes during which fusion may begin to occur. At the end of this time, if no fusion has occurred, add several drops of medium to the slide and view again.

Results

1. Draw a picture showing: colorless and red protoplasts, fusion products.
2. Label the following: plasma membrane, vacuole, nucleus.

3. How can you distinguish between fused and non-fused protoplasts?
4. If a problem arises, it is likely to be that protoplast yield or yield of fused protoplasts is low. Even under these circumstances, positive results from at least a portion of the class should be forthcoming from the experiment.
5. Since handling the protoplasts under sterile conditions, as would be done in a research lab, presents difficulties for the teaching lab, the experimental protocol ends with the production of fusion products. Students should understand how the experiment would be performed differently if the goals of the experiment were to maintain the hybrids in culture. This is discussed in the student guide.
6. The plant material used in the experiment, red onion, was chosen because it is readily available and produces a good yield of protoplasts within the time limitations imposed by a laboratory class. The other benefit of using red onion for this experiment is that digestion of the red onion tissue gives a mixed population of cells, red and colorless. This means that for the fusion experiment included in the module, it is not necessary to prepare two different plant tissues. Other plant materials may perform equally well, but substitutions should be tested before class since all plant tissues do not respond equally to this mixture of enzymes and osmoticum. Some tissues will take from a few hours to overnight to release a sufficient number of protoplasts for the fusion experiment. If desired, protoplasts may be prepared before class and used for the fusion experiment alone during the laboratory period.

7

MAINTENANCE OF A CELL LINE

To culture and maintain established cell lines, a primary culture is split to produce new cultures and is subsequently known as a cell line. The two types of cell lines are (a) finite cell line and (b) continuous or established cell lines. A finite cell line is capable of a limited number of cell generations in vitro and after which cells die. A continuous cell line has the capacity for an infinite number of population doubling and thus in a sense immortal. Some continuous cell lines are also malignant, meaning they will grow as a tumor and invade other tissues if injected into an animal.

Earle is credited with establishing the first immortal cell lines in 1940. Most of these were established by treating rodent primary culture with chemical agents. One of these lines called L-cells was established in 1943 by treating primary mouse fibroblasts with the cancer causing agent methylcholanthrene. George and Margaret Gey derived cells from the cervical carcinoma of a black woman named Henrietta Lacks. This cell line which is called "HeLa", is the most common source of human cells used in research today. Today, many cell lines are available which can be obtained from American Type Culture Collection (ATCC), local medical colleges or any research institutions.

Established cell lines can be divided into two groups: those that can grow attached to a solid surface (monolayer) and those that grow in suspension. The mouse L-cells is a good example of the surface dependent cell line. When these cells are seeded in a culture vessel, they attach and grow to form a uniform layer of cells or a monolayer, which is always one cell thick. These cells will stop growing once they start filling up the surface and touching each other due to contact inhibition. In order to subculture a monolayer, it has to be dissociated to individual cells with trypsin, after which it can be put in a larger vessel with more space and fresh medium. Suspension cells are grown in liquid culture similar to bacterial culture.

Safety Guidelines

Wear gloves, lab coat, and safety glasses. Standard laboratory safety procedures should be followed.

Experimental Outline

- Maintain an established cell line in tissue culture
- Introduce Trypsin-EDTA into the culture flask, aseptically to help maintain conditions for a monolayer culture
- Observe individual cells using an inverted microscope
- Maintain cells, replace medium when a drop in the pH is observed based on change on phenol red indicator - changing from red to yellow

Materials

Chick embryo fibroblasts (from Module 11) or mouse L-cells

A bottle of PBS

A bottle of 1x Trypsin-EDT A solution (concentrated)

Trypan blue solution - 0.5% in PBS

A bottle of complete medium (HMEM), supplemented at 10% v/v with Fetal Bovine Serum and 1% v/v penicillin-streptomycin 100x stock

Waste container

10 ml pipets, sterile and pipet pump

CO_2 incubator

Method

1. Keep PBS, Trypsin-EDTA and complete medium in a 37°C water bath.
2. Observe the condition of the monolayer by holding the bottle up to the light. In a good culture the medium will appear clear and the cells will look smooth and confluent with no clumps.
3. Observe the individual cells by using an inverted microscope at low power. Only those cells which undergo mitosis would be round and floating.
4. Make sure that the monolayer is confluent and if so, it is ready to be transferred.
5. Uncap culture flask and decant medium to a waste container.

6. Aseptically withdraw 10 ml of sterile PBS, and transfer it into the culture flask and recap the flask. Rock the flask gently from side to side for 10-15 seconds to rinse the cells. This step helps to remove the serum, which would inhibit the enzymatic activity of trypsin. Decant the solution into a waste container.
7. Add 0.2 to 0.5 ml of Trypsin-EDTA into the culture flask, swirl the trypsin around and incubate at 37°C for 3 minutes. Too much trypsin can damage cells. Cells are attached to cells and substrate by surface glycoproteins, serum proteins and chemical groups that coat the tissue culture-treated plastic surfaces of the flask. The enzyme trypsin catalyzes the hydrolysis of bonds involved in cell to cell and cell to substrate contact.
8. Once the cells begin to dislodge, tap the flask against the palm of your hand lightly. Hold the flask up to the light to check that no patches of monolayer remain. Observe the cells under an inverted microscope to confirm their presence.
9. Aseptically add 5 ml of complete medium. Proteins in the serum will stop the action of trypsin.
10. Aspirate the cell suspension by drawing it up into the pipette and expelling it back into the flask 5-6 times with the tip of the pipette pressed against the bottom corner of the flask. This will break up the cell clumps. Clumped cells will affect the accuracy of the cell count.
11. Count the cells using a hemacytometer.
12. Expand the culture by taking 1×10^5/ml for chicken fibroblasts or 1×10^4/ml for L-cells. If you are using 25 cm^2 flasks, transfer about 1 ml of cell suspension to each flask and add 9 ml of medium to the flask.
13. Incubate the culture at 37°C for 12-24 hours. After this, decant the old medium and add 10 ml of fresh medium and incubate. It is necessary to transfer (split) these cultures once they become confluent, which should occur in 3-4 days. Observe them under inverted microscope every 2-3 days. When the phenol red indicator in the medium changes from red to orange-yellow or yellow which indicates a drop In pH, decant and discard the medium and replace it with fresh medium.

Results

Results can vary from group to group. Students can quantify viable cells when utilizing the inverted microscope.

8

Isolation of Chloroplasts

Chloroplasts are the subcellular sites of photosynthesis, the process by which green plants, using energy from light, produce carbohydrate and oxygen from carbon dioxide and water. Under the microscope, chloroplasts are recognized as bean-shaped, membrane-bound, green (chlorophyll-containing) organelles. In this experiment, chloroplasts will be isolated from a cell homogenate by density gradient centrifugation. A measurement of the amount of chlorophyll in the preparation will be used to assess yield of intact chloroplasts. Finally, the chloroplasts will be saved for the isolation of DNA.

Safety Guidelines

Follow standard laboratory safety practices.

Experimental Outline

Prepare sucrose gradients (30 min)
Harvest pea tissue and process sample (30 min)
Apply sample to gradient and centrifuge (30 min)
Retrieve chloroplasts from gradient and assay yield (30 min)

Materials

Pea seedlings
Homogenization buffer
10 mM KCl
1 mM $MgCl_2$
1 % (w/v) Dextran T40
1 % (w/v) Ficoll
0.1% (w/v) Bovine Serum Albumin

Make to volume with 30% (w/w) sucrose in 0.1 M Tricine buffer, pH 7.5

Ice bucket

Cheesecloth and Miracloth (Calbiochem)

Centrifuge tubes

Sucrose (gradient) solutions (w/w, prepared in 0.1 M Tricine buffer, pH 7.5) 60%, 50%, 40%, 30%

Pasteur pipettes

Blender

Centrifuges (clinical and super speed)

Spectrophotometer

Pre-lab Preparation

Timetable of events

The exercise can be divided into four parts: (1) preparation of sucrose gradients, (2) sample preparation, (3) centrifugation, and (4) analysis of results. Each of these activities will require approximately 30 minutes.

Equipment requirements

The protocol described here is written to be used with a refrigerated, super-speed centrifuge (e.g. Sorvall RC series) equipped with a swinging bucket rotor (e.g. Sorvall HB4).

Other comparable centrifuges will perform equally well, but specified speeds and times of centrifugation might have to be changed slightly. The operations manuals for your particular centrifuge and rotor will explain these changes.

The assay of chlorophyll content in gradient fractions requires the use of a spectrophotometer or colorimeter.

A blender is required for tissue homogenization. A standard household blender will work well, but the blades should be as sharp as possible (some laboratories have been known to modify their blenders by replacing the standard blades with razor blades!).

Plant Material: The protocols in this unit all call for the use of pea seedlings as starting material. The seed is relatively inexpensive and easy to grow. For best results, use fresh seed. Seedlings should be about seven days old for the lab. To plant, soak seed overnight in a large container. Sow the seeds on a layer of about 1.5 inches of wet horticultural grade vermiculite in a standard nursery flat (21 × 10 × 2 inches). The seed can be sown thickly nearly touching one another.

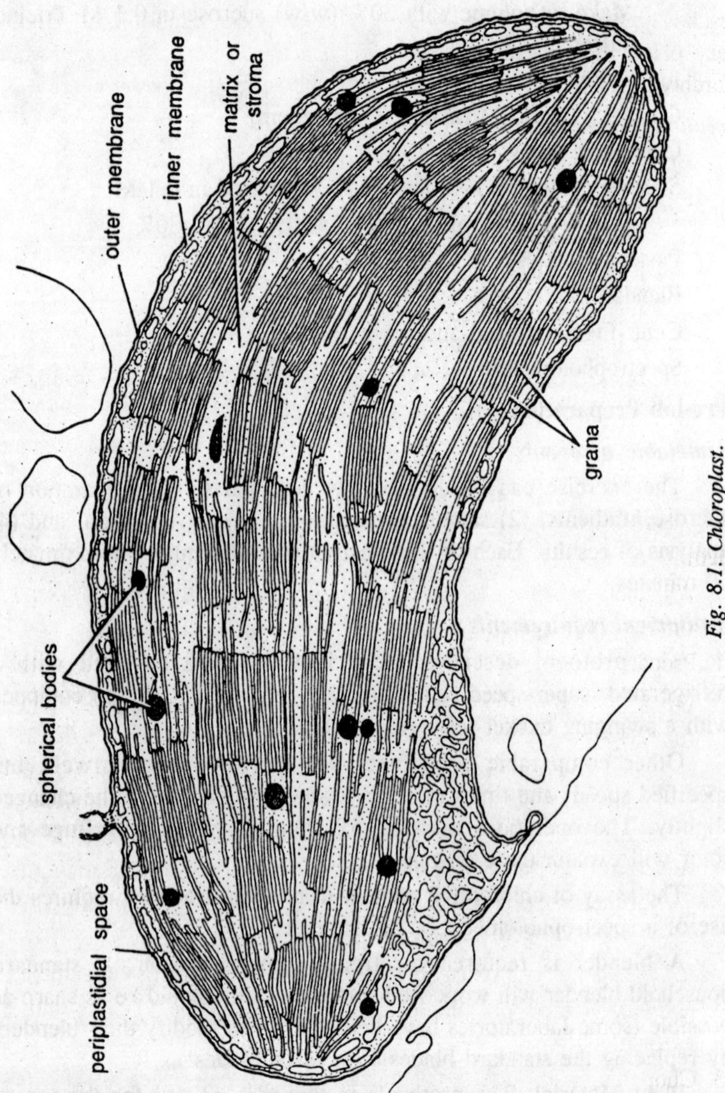

Fig. 8.1. Chloroplast.

(We usually use a 500 ml beaker-full of dry seed to plant one flat. At harvest, you can expect about 300 g of shoot tissue from such a planting.) Cover the seed with 0.5-1 inch of vermiculite and water well. Cover the flat with plastic wrap to hold in moisture until the seedlings begin to emerge. Once the seedlings have emerged and the plastic has been removed, keep well-watered. The seedlings can be grown in the lab on a window sill, in a growth chamber or in the greenhouse. For best results, however, do not grow the seedlings under intense light. Under very bright lights, chloroplasts tend to accumulate large granules of starch and these can do damage during blending and centrifugation.

Solutions required

Homogenization Buffer

10 mM KCI
1 mM $MgCl_2$
1 % (w/v) Dextran T40
1 % (w/v) Ficoll
0.1 % (w/v) Bovine Serum Albumin
Make to volume with 30% (w/w) sucrose in 0.1M Tricine buffer, pH 7.5

Sucrose (gradient) Solutions

60%,50%,40%, and 30% sucrose (w/w) in 0.1M Tricine buffer, pH 7.5

Method

All solutions should be ice-cold. Keep solutions, samples, gradients, etc. on ice while you are working.

1. Prepare two sucrose gradients in 50 ml centrifuge tubes.
   (a). Pipette 5 ml 60% sucrose solution into bottom of each tube.
   (b). Layer 5 ml 50% sucrose, then 10 ml 40% sucrose into each tube. Layers should be distinct from one another if you are careful. (Hint: Tip the tube as you add each layer of sucrose. Let tip of the pipette just touch the surface of liquid in tube.)
   (c). With the tip of a pasteur pipette or stirring rod, gently mix at the interface of the 50% and 40% layers to diffuse slightly.
   (d). Layer 5 ml 30% sucrose on top of the gradient.
   (e). Keep the gradients on ice while you prepare tissue sample.
2. Harvest 5 grams of 7-day-old pea seedlings at the soil line with a razor blade.
3. Chop the tissue into small pieces with a razor blade or scissors and transfer them to a chilled blender containing 20 ml ice-cold homogenization buffer. (Note: More than one 5 g batch of seedlings can be blended at a time. If several lab groups are sharing the blender, they should use 20 ml of buffer for each 5 g of seedlings homogenized, then divide the homogenate.)

4. Homogenize with five 2-3 second bursts of the blender at high speed.
5. Filter the homogenate into a beaker (on ice) through four layers of cheesecloth, squeezing the cloth gently to remove most of the liquid (wear gloves).
6. Re-filter the first filtrate through one layer of Miracloth, moistened in homogenization buffer, by gravity. Do not squeeze. You may want to prepare a wet-mount slide of the residue left in the cheesecloth or Miracloth for the microscope. What have you removed from the homogenate by filtration?
7. Layer 10 ml of the filtrate onto the top of each of your gradients (prepared in step 1). Check to see that the two gradients are balanced against one another. If necessary, add homogenization buffer to make the tubes balance. Centrifuge at 4°C in HB4 rotor, 4000 rpm for 5 minutes, then increase speed to 10,000 rpm for 10 minutes. Allow the centrifuge to coast to a stop. Carefully remove your gradients from the rotor.
8. You should see two green bands in the gradient. The green band toward the bottom of the tube is the fraction containing intact chloroplasts. Remove the top of the gradient carefully with a pasteur pipet. Save the two chlorophyll-containing fractions in clean tubes on ice.
9. Prepare wet-mount slides of the chlorophyll-containing fractions and examine on the microscope. What differences do you notice between them?
10. Assay chlorophyll content of the two green bands. For each sample:
    (a) Into a clean centrifuge tube. pipette 50 μL of the gradient fraction to be assayed and 0.95 ml distilled water.
    (b) Add 4 ml acetone.
    (c) Centrifuge in a clinical centrifuge, 5 minutes.
    (d) Measure the absorbance of the solution in a spectrophotometer at 652 nm. (The appropriate blank for this measurement is 80% acetone in water.)
    (e) Calculate chlorophyll content:
    $A_{652} \times 29 = \mu g$ chlorophyll/10 μl chloroplast fraction.
11. Freeze the intact chloroplast fraction to save for DNA isolation.

Results

Make a diagram of your gradient. Have materials other than chloroplasts banded in the tube? Where are they? What do they look like? Why was the cell homogenate filtered before being loaded on the gradient? What was removed by filtration? What fraction of the chloroplasts in the homogenate have you isolated intact?

After centrifugation, the gradients should show two bands of chlorophyll. The upper of the two contains broken chloroplasts and membrane fragments. The lower of the bands (about three quarters of the distance to the bottom of the tube) contains intact chloroplasts. Probably the most frequent cause of a poor yield of intact chloroplasts is over-blending. Using any type of homogenizer to disrupt tissue is a trade-off between efficiency in breaking cells open and generating so much shear in the solution that organelles are also disrupted.

If you plan to use the chloroplasts from this prep for DNA isolation avoid contaminating the intact chloroplast fraction with broken chloroplasts. Also, try to retrieve the intact chloroplast fraction from the gradient in as small a volume as possible.

9

ISOLATION OF GENOMIC DNA

Isolation of DNA from plant tissues is at the heart of plant molecular biology. Because plant cells are surrounded by rigid cell walls and because plant tissues often contain a variety of secondary metabolites that can damage DNA. Thus, DNA isolation from plants can present some particular difficulties. Among the many protocols developed for DNA isolation from plants, the method presented here is one of the simplest and most effective for a variety of plant species.

DNA isolated by methods such as the one presented here represents total cellular DNA. In plants, this means the isolation of three distinct genomes:

1. The nuclear genome.
2. The chloroplast genome.
3. The mitochondrial genome.

Generally, when we talk about plant DNA we mean nuclear DNA, but it is important to remember that chloroplasts and mitochondria each have distinct genomes. These genomes, like the nuclear genome, are targets of research in molecular biology and genetic engineering. Total plant DNA isolated here will be compared with DNA isolated from chloroplasts in another laboratory exercise.

Safety Guidelines

Reagents used in this protocol are potentially dangerous. CTAB, cetyltrimethylammonium bromide (in the extraction buffer) is a strong detergent and can cause burns to the skin. Chloroform is toxic by inhalation or on contact with skin. Follow your instructor's direction for proper handling and disposal of these materials.

Experimental Outline

Day 1. DNA isolation (2 hrs)
Grind tissue and suspend in extraction buffer (EB)
Incubate at 65°C for 1 hr
Extract with chloroform
Precipitate DNA with isopropyl alcohol
Resuspend the DNA in buffer and store

Day 2. Compare this DNA preparation with DNA isolated from chloroplasts by gel electrophoresis.

Materials

EB (extraction buffer): 50 mM Tris pH 8.0,1% CTAB, 50 mM EDTA.1 mM 1,10-O-phenanthroline. 0.7 M NaCl, 1% betamercaptoethanol
Chloroform
Isopropyl alcohol
80% ethanol, 15 mM ammonium acetate, pH7.5
TE buffer: 10 mM Tris pH 8.0, 1 mM EDTA
Centrifuge Tubes (50 ml capacity, capped)
Water bath 65°C
Centrifuge

Pre-lab Preparation

Pea tissue used in this protocol should be air dried or freeze dried. Grow the plants according to directions. Tissue can be air dried in the laboratory by spreading the cut shoots on a sheet of absorbent paper and turning daily. Tissue treated in this way will typically dry in 5-7 days. Dried tissue can be saved for use at a later date by sealing in a tightly closed jar and freezing.

Method

1. Grind 1 gram dried (air-dried or freeze-dried) pea shoots to a fine powder in a mortar with pestle.
2. Mix the powder with 25 ml EB (extraction buffer) in a 50 ml, capped centrifuge tube.
3. Place the tube into a 65°C water bath and incubate it there for 1 hr. Several times during the hour, mix the tube's contents by inversion.
4. Remove the tube from the water bath and allow it to cool for several minutes on the bench. (Note: Do not skip this step. The

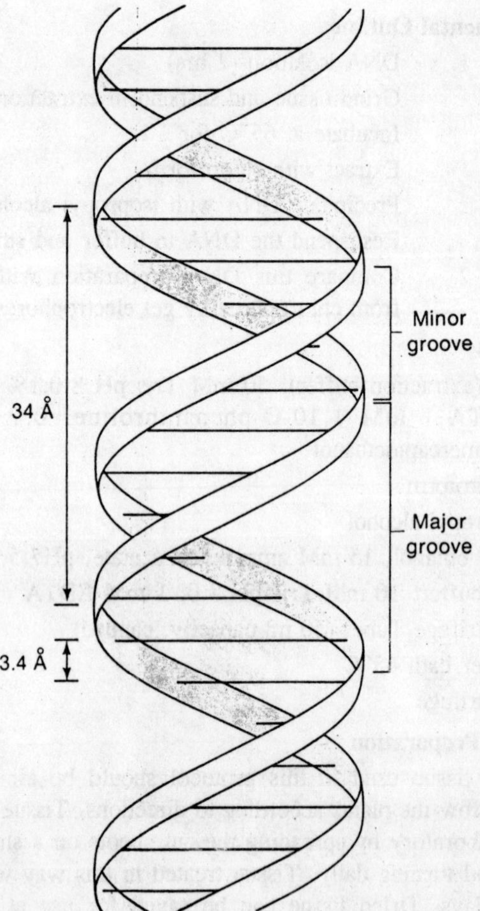

Fig. 9.1. Diagrammatic model of the DNA double helix in the common B form. In the A form, which may not exist within cells, the bases are tilted and there are 11 bases per turn.

chloroform added in the next step will boil out of the tube if added at 65°C.)

5. Add 20 ml chloroform to the tube, cap and mix by inversion until the contents are thoroughly mixed. When mixed, the extraction buffer and the chloroform will form a thick emulsion.
6. Centrifuge the tube at > 3500 x g for 10 minutes to break the emulsion and separate the tube contents into two phases.
7. Upon removal from the centrifuge, the contents of the tube form three distinct layers. At the bottom is a green layer of chloroform.

Often, bits of plant material are pelleted under this layer. In the middle is an interphase consisting largely of denatured protein and bits of leaf tissue. This layer is whitish or yellowish in color. At the top of the tube is the straw-colored aqueous layer. This top layer contains the majority of the DNA in the preparation. Using a pipette, remove this top layer to a small flask or beaker. Avoid transferring any of the interphase material from the centrifuge tube.

8. What remains in the centrifuge tube, that is, interphase and organic (chloroform) layers are hazardous waste. Follow your instructor's direction for proper disposal.
9. To the aqueous phase in the flask or beaker, add 2/3 volume of isopropyl alcohol (for example, if you have transferred 24 ml to the flask, add 2/3 x 24 or 16 ml of isopropyl alcohol). Mix by swirling the contents. DNA will precipitate to from a cottony mass.
10. Using a glass rod or a pasteur pipet, transfer the DNA to a clean flask. Add 10 ml of 80 % ethanol, 15 mM ammonium acetate and swirl to wash.
11. After about 20 minutes, transfer the precipitated DNA to a microcentrifuge tube and centrifuge briefly to drive the DNA to the bottom of the tube. Using a pasteur or capillary pipet, remove the residual ethanol. Allow the DNA to dry in the uncapped tube for about 10 minutes on the bench top.
12. Add 0.75 ml of TE buffer to the DNA to dissolve the precipitate. (Note: Large quantities of DNA may require some time to dissolve completely. Leave the capped tube in the refrigerator until the next lab class.)

Day Two:

Cut 2 μl of the DNA with a restriction endonuclease as directed by your instructor.

On an agarose gel, compare total DNA (both uncut and cut) with DNA isolated from chloroplasts (both uncut and cut).

Results

The protocol is generally trouble-free. The only difficulties we have experienced with it come from: (1) using incompletely dried material, and (2) using heat-dried material. Rarely, DNA prepared by this method forms a flocculent rather than cottony precipitate. Such precipitates will not spool on a glass rod and must be collected by centrifugation. If desired, RNA can be removed from the DNA preparation by adding 20 micrograms of RNase A (heat-treated to destroy DNases) along with the TE buffer in Step 12 of the protocol.

Note on the use of chloroform in this exercise: If you want to avoid the use of chloroform in this exercise, make the following modifications to the lab protocol:

1. Grind tissue as before, mix with EB, and incubate in the water bath.
2. Centrifuge to pellet undigested plant tissue.
3. Transfer the supernatant to a fresh tube, flask or beaker and precipitate DNA with isopropanol as before. Note: The supernatant will be colored dark green in this preparation making the precipitation of DNA somewhat more difficult to see. If this is a problem, the supernatant can be diluted with additional EB or TE buffer before the precipitation step. DNA prepared in this way should not be expected to cut well with restriction endonucleases, but it may be satisfactory.

10

Isolation of Chloroplast DNA

The DNA of plant cells is found in three distinct genomes. First, there is nuclear DNA, familiar as the DNA that makes up the chromosomes. But mitochondria and chloroplasts each have DNAs of their own. These genomes are closed circular DNA molecules encoding many of the enzymes necessary for the function of the organelles. Because of the importance of mitochondria and chloroplasts to the cell, their DNA is of interest to molecular biologists and biotechnologists. The chloroplast DNA of several species of plants have been cloned and sequenced in their entirety. In at least one organism (the green alga Chlorella), chloroplast as well as nuclear genomes have been genetically transformed. In this laboratory exercise, DNA will be isolated from chloroplasts and compared with total DNA. The objectives of this exercise are:

1. To isolate DNA from chloroplasts, and
2. To compare chloroplast DNA to genomic DNA.

Safety Guidelines

Some reagents used in this protocol are potential hazards. CTAB, cetyl-trimethylammonium bromide, is a strong detergent and can cause burns to the skin. Chloroform is toxic by inhalation or contact with skin. Follow instructor's direction for proper handling/ disposal of these materials.

Experimental Outline

Day 1 DNA isolation (2 hrs.)
Add EB buffer to chloroplast prep
Incubate 1 hr, 65°C

Extract with chloroform
Precipitate DNA
Resuspend in Buffer

Day 2 Compare DNA from chloroplast to genomic DNA.
Restriction Digest of DNA (optional)
Pour gel, load cut and uncut DNA's

Materials

EB (extraction buffer): 50 mM Tris pH 8.0,1% CTAB, 50 mM EDTA, 1 mM 1,10-O-phenanthroline, 0.7 M NaCI, 0.1% beta-mercaptoethanol
Chloroform
Isopropyl alcohol
TE buffer: 10 mM Tris pH 8.0,1 mM EDTA
Centrifuge Tubes
Water Bath (65°C)
Centrifuge

Pre-lab Preparation

No special preparation beyond the requirements for those exercises is required for this activity.

Method

1. Start with the frozen chloroplast preparation. Typically, this sample will have a volume of several milliliters. For each milliliter of chloroplasts, add 4 ml EB. If necessary, transfer the mixture to a capped centrifuge tube of at least twice the volume of the chloroplasts and EB.
2. Incubate the mixture at 65°C for 1 hr.
3. Remove the tube from the water bath and allow to cool on the bench top for several minutes before proceeding.
4. Add an approximately equal volume of chloroform to the tube, recap and mix by inversion.
5. Centrifuge the tube at > 3500 x g for 10 minutes.
6. Upon its removal from the centrifuge, the tube contents will have separated into two distinct layers. Using a pipette, transfer the upper (aqueous) layer into a fresh centrifuge tube. (This tube should be of the same size as that used in the first step.) The lower (organic) layer is hazardous waste. Follow your instructor's directions for proper disposal.

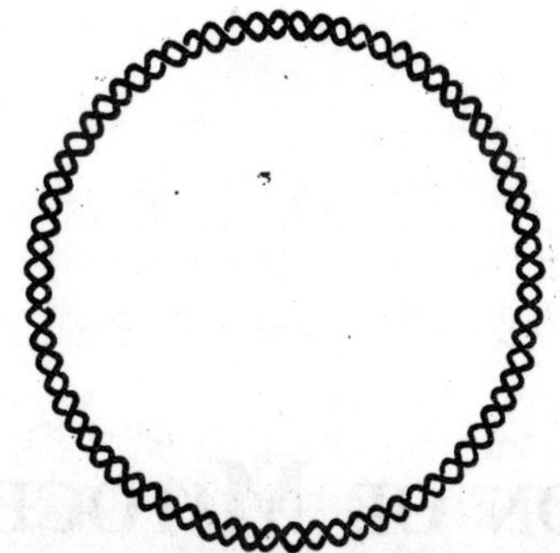

Fig. 10.1. A nonsupercoiled covalent circle having 36 turns of the helix.

7. Add 0.6 ml of isopropanol for each ml of DNA-containing extract in the centrifuge tube. Mix by inversion.
8. Centrifuge at > 10,000 x g for 20 minutes.
9. After centrifugation, decant the liquid in the tube away from the DNA-containing pellet. Stand the tube upside down on a paper towel or "Kimwipe" for several minutes to allow the liquid to drain. The tube's inside can be wiped carefully to remove liquid, but take care not to dislodge the DNA pellet.

Results

The yield of chloroplast DNA is expected to be low and will depend in some measure on the quality of the chloroplast preparation produced by the students. Also note that chloroplast DNA may not cut well with restriction endonucleases. Still, it is likely that at least one lab group or individual will get results that can be shared with the rest of the class; Some of the students will accept this activity as a challenge. Since it is teamed with an examination of total DNA, none of the students will be completely without results. The exercise is worth a try if for no other reason than it will give the students some experience in working with very small quantities of DNA.

11

Isolation of Mitochondria

In the following experiments, you will isolate mitochondria from the roots of pea seedlings by differential centrifugation and assay the activity of a mitochondrial enzyme to assess the success of the isolation protocol. Recall that the technique of differential centrifugation separates cell components based on differences in the rate at which they sediment in a centrifugal field. In the protocol used here, a first centrifugation step will remove whole cells, cell wall fragments, nuclei, starch, etc. Mitochondria, because of their small size will not be pelleted by this step, but will remain suspended in the supernatant. A subsequent spin at a higher speed will then be used to pellet the mitochondria to the bottom of the centrifuge tube.

To assess the efficiency of the isolation protocol, you will assay the activity of a "marker enzyme" in the mitochondrial pellet and in the supernatant from which the mitochondria are isolated. A marker enzyme is any enzyme whose activity is confined to the organelle being isolated. If the isolation protocol is 100% effective, all marker enzyme activity should appear in the mitochondrial fraction. The marker enzyme to be assayed in this experiment is cytochrome c oxidase.

Enzyme activity is measured by determining the rate of oxidation of cytochrome c. This can be followed, using a spectrophotometer, by an increase in absorbance at 550 nm by the reaction mixture during the initial, linear phase of the reaction.

Safety Guidelines

Follow standard laboratory safety procedures.

Experimental Outline

Harvest root tissue

Prepare extract
Isolate mitochondria by differential centrifugation
Assay mitochondrial activity

Materials

Method 1 & 2

- Pea seedlings
- Ice bucket
- Homogenization buffer
  - 70 mM sucrose
  - 220 mM mannitol
  - 0.5 g/l Bovine Serum Albumin
  - 2.0 mM HEPES pH 7.4
- Cheesecloth/Miracloth
- Centrifuge tubes
- Pasteur pipets
- Small paint brush
- Blender
- Centrifuge (super speed)
- Spectrophotometer

Method 3

- Pea seedlings (grow seedlings as for regular protocol, but use whole seedlings for this experiment)
- Homogenization buffer (same as in regular protocol)
- Cheesecloth/Miracloth
- Centrifuge tubes (4 per lab group)
- Test tubes (3 per lab group)
- Pasteur pipets
- Dropper bottles of:
  - Methylene blue
  - Janus green
  - Iodine (KI)
  - Vegetable or mineral oil
- Clinical Centrifuge
- Blender
- 37°C Water Bath

Pre-lab Preparation

Equipment requirements

The protocol described here is written to be used with a refrigerated, super-speed centrifuge equipped with a fixed angle rotor (e.g. Sorvall SS34).

Other comparable centrifuges will perform equally well, but specified speeds and times of centrifugation might have to be changed slightly. The operations manuals for your particular centrifuge and rotor will explain these changes. If a super-speed centrifuge is not available, consider the alternative protocol for cell fractionation described below.

The assay of mitochondrial marker enzyme activity requires the use of a spectrophotometer with a band width of 5 nm or less. Before using the cytochrome c solution in the assay, check to see that it is completely reduced. To do this, add 1 or 2 crystals of sodium dithionate to 1 ml of cytochrome c stock solution. Transfer solution to the spectrophotometer cuvette and measure its absorbance at 550 nm and 565 nm. The A_{550}/A_{565} ratio should be 9-10.

A blender is required for tissue homogenization. A standard household blender will work well, but the blades should be as sharp as possible (some laboratories have been known to modif, their blenders by replacing the standard blades with razor blades!).

Plant material

The protocols in this unit all call for the use of pea seedlings as starting material. The seed is relatively inexpensive and easy to grow. For best results, use fresh seed. Seedlings should be about seven days old for the lab. To plant, soak seed overnight in a large container. Sow the seeds on a layer of about 1.5 inches of wet horticultural grade vermiculite in a standard nursery flat (21 × 10 × 2 inches). The seed can be sown thickly - nearly touching one another. (We usually use a 500 ml beaker-full of dry seed to plant one flat. At harvest, you can expect about 300 g of root material.) Cover the seed with 0.5-1 inch of vermiculite and water well. Cover the flat with plastic wrap to hold in moisture until the seedlings begin to emerge. Once the seedlings have emerged and the plastic has been removed, keep well-watered. The seedlings can be grown in the lab on a window sill, in a growth chamber or in the greenhouse. Shoot material, not used in this exercise should be saved, air dried or freeze dried, and stored for genomic DNA isolation.

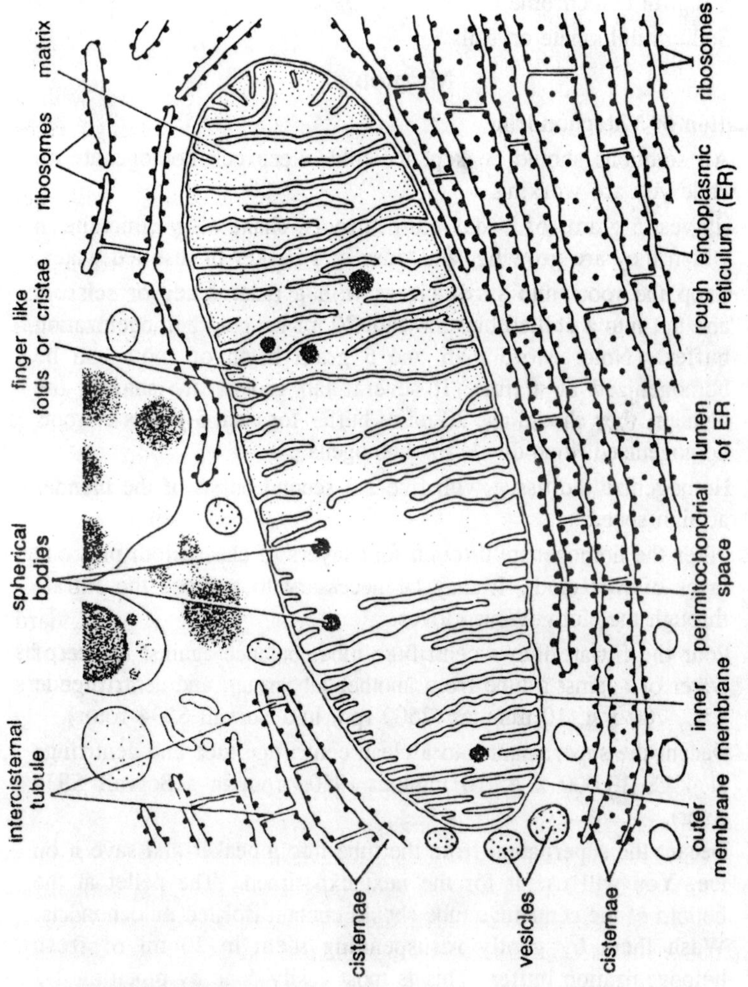

Fig. 11.1. Mitochondria and endoplasmic reticulum.

Solutions required

Homogenization Buffer

70 mM sucrose

220 mM mannitol

0.5 g/l Bovine Serum Albumin

2.0 mM HEPES pH 7.4

0.1 M potassium phosphate buffer, pH 7.4

0.8M ascorbic acid

4% triton X-100

5 mg/ml cytochrome c
Sodium dithionate crystals

Method

Isolation of Mitochondria

All solutions should be kept ice-cold. Keep cell homogenate on ice while you are working.

1. Harvest 5 grams of 7 -day-old pea roots, shake off vermiculite in which they are growing, and rinse in a beaker of distilled water.
2. Chop the roots into small pieces with a razor blade or scissors and put into a chilled blender with 20 ml ice-cold homogenization buffer. (Note: More than one 5 gram batch of roots can be homogenized at a time. If several lab groups are sharing the blender, they should use 20 ml of buffer for each 5 grams of roots homogenized, then divide the homogenate.
3. Homogenize the tissue with five 2-3 second bursts of the blender at high speed.
4. Filter the homogenate through four layers of cheesecloth plus one layer of Miracloth. It may be necessary to squeeze the filtrate through the cloth. Wear gloves.
5. Pour the filtrate into a centrifuge tube, balance against a tube of water or against a tube from another lab group, and centrifuge at 4°C, 700 x g, 10 minutes (2500 rpm in a Sorvall SS34 rotor).
6. Decant the supernatant into a clean centrifuge tube and centrifuge at 4°C, 10,000 x g, 10 minutes (9500 rpm in a Sorvall SS34 rotor).
7. Decant the supernatant from the tube into a beaker and save it on ice. You will use it for the next experiment. The pellet at the bottom of the centrifuge tube should contain isolated mitochondria. Wash them by gently resuspending them in 20 ml of fresh homogenization buffer. This is most easily done by pipetting 1-2 ml of the buffer into the tube and using a small paint brush to break up the pellet. Once the pellet is resuspended In this small volume, it can be diluted with the remaining 18-19 ml of buffer.
8. Recentrifuge the washed mitochondria as in step 6 above.
9. Discard the supernatant from this spin and resuspend the mitochondrial pellet in 5 ml of homogenization buffer.

Assay of a Mitochondrial Marker Enzyme

1. From the ice bucket, remove 0.2 ml of the mitochondria preparation to each of two small test tubes and 0.2 ml of the

10,000 x g supernatant to one small test tube and allow them to come to room temperature on the lab bench. To one of the tubes containing mitochondria, add 0.8 ml of potassium phosphate buffer (a 1/5 dilution). To the other tube of mitochondria, add 1.8 ml buffer (a 1/10 dilution).

2. Assemble four reaction mixtures in four small test tubes:

Reaction Mixture (in each test tube)

0.1 M potassium phosphate buffer, pH 7.4	50 μl
4% Triton X-100	25 μl
5 mg/ml cytochrome c	50 μl
Distilled water	675 μl
Sample to be assayed (mitochondria dilutions or supernatant)	200 μl

3. One of the four tubes above contains no sample to be assayed. Transfer the contents of this tube to a cuvette and use it as a blank in the spectrophotometer. If your spectrophotometer requires a reference cuvette, blank the cuvettes with water and use the contents of this tuba as your reference.
4. To perform the assay, transfer a reaction mix containing sample to a cuvette and place into the spectrophotometer. Record the absorbance at 550 nm at 20 second intervals for one minute to determine the rate of the reaction in the absence of substrate. This value (the slope of this line) will be used in step 7 below as a correction factor in your calculation of the rate of the enzyme-catalyzed reaction. If your spectrophotometer uses a reference cuvette, this step is not necessary.
5. Start the reaction by removing the cuvette from the spectrophotometer and adding 10 μl of 0.8M ascorbic acid. Mix contents of the cuvette by Inversion and quickly replace the cuvette in the spectrophotometer. Record the absorbance at 550 nm at 20 second intervals for 2 minutes.
6. Repeat steps 4 and 5 for each sample to be assayed.
7. Calculate the rate of cytochrome c oxidation by each of the samples:

$$\text{Rate} = \frac{\text{Change in absorbance / minute} - \text{Change in absorbance / minute}}{\text{Molar absorptivity of cytochrome} \times \text{path length of light through the cuvette}}$$

Molar absorptivity of cytochrome c = 18.5 x 10^6 M^{-1} Path length through cuvette is usually 1 cm.

The units attached to rate are moles of cytochrome c oxidized per minute. In general, this value would be reported as micromoles cytochrome c oxidized per minute, thus, calculation is simplified to:

$$\text{Rate} = \frac{\text{Change in absorbance / minute (corrected as above)}}{18.5}$$

Note: Enzyme activity is usually reported In the literature as specific activity. This is a way of standardizing the reporting of enzyme activity since the exact molar concentration of enzyme in the preparation being assayed is not known. How is specific activity related to the rate of reaction just calculated? Specific activity = rate/ mg protein (total) in the assay mixture.

Protocol Using Clinical Centrifuge Only

Objectives

1. Separate a cell homogenate into several fractions by differential centrifugation.
2. Compare the composition of the fractions by microscopy.
3. Assay mitochondrial activity in the fractions by a qualitative assay.

All solutions should be ice-cold. Keep samples on ice at the lab bench if possible.

1. Harvest whole pea seedlings and wash in distilled water to remove planting material. Blot dry with paper towels.
2. Weigh out about 50 grams of seedlings, chop them into small pieces with a razor blade or scissors and transfer them to a chilled blender. Add 250 ml ice-cold homogenization buffer.
3. Homogenize tissue with five, 2-3 second burst of the blender at high speed.
4. Divide homogenate among lab groups. Each group should have 20-25 ml (enough for two, 10-15 ml centrifuge tubes).
5. Filter the homogenate through four layers of cheesecloth plus one layer of miracloth into a beaker on ice. Gently squeeze the cloth to remove most of the liquid. Wear gloves. Save the residue in the cheesecloth for examination later. Note: Homogenization of tissue and filtration step can be done prior to class by the instructor.
6. Divide filtrate in 2 centrifuge tubes. Tubes should be filled to same level to keep centrifuge balanced. Label tubes 1 and 2.

7. Centrifuge for three minutes at 200 x g . Start timing when the centrifuge rotor reaches top speed. Allow the rotor to coast to a stop when three minutes are up.
8. Remove the tubes from the rotor. Decant the supernatant from tube 2 into a clean tube (label the new tube 3). Save the pellet in tube 2 on ice. Add buffer to tube 3 to balance it against tube 1.
9. Return the tubes (1 and 3) to the centrifuge and spin at 700 x g for 10 minutes.
10. Remove the tubes from the rotor.
11. Decant the supernatant from tube 3 into a fresh tube (label 4). Save the pelleted material in tube 3.
12. Prepare wet-mount slides of the cell fractions (residue In cheesecloth and tubes 1-4) and examine them on the microscope. The residue in the cheesecloth consists largely of unbroken pieces of seedling tissue, cell debris, etc. Tube 2 contains material pelleted at low speed (200 x g) from the homogenate. Can you identify any of these components? Add a drop of iodine solution to the edge of the coverslip of your slide. Any starch grains present should turn blue-black in the presence of iodine. The pellet in tube 3 was separated from the homogenate at higher speed (700 x g) and for a longer time (10 minutes) than that in tube 2. How is this pellet different from the pellet in tube 2? Are any of the components present in these two pellets the same? The supernatant in tube 4 should be relatively clear. Only the smallest cell components remain in this solution. Prepare a wet-mount slide of this supernatant. Add a drop of Janus green stain. With the stain, at high magnification, mitochondria appear as tiny, dark specks. Examine tube 1. This tube shows a history of the entire experiment and illustrates the principle behind differential centrifugation. The sediments at the bottom of the tube are arranged in layers. The largest, most dense cell components are at the tube's bottom. Additional layers of sediment were added to the pellet by higher centrifugal forces applied for longer periods of time. Only the tiniest of particles remain in solution at the end of the experiment.
13. Assay the activity of mitochondria. To demonstrate that mitochondria remain in the supernatant of the high speed spin (tube 4) and that they have been separated efficiently from other cell components, perform the following experiment:

As you know, mitochondria are the subcellular sites of respiration. The activities of these organelles thus consume oxygen. Methylene blue dye is blue in the presence of oxygen, but is colorless when reduced (i.e. when oxygen is removed from it). In this experiment, the presence of mitochondria is detected by the disappearance of blue color from the reaction mixture. Prepare 3 test tubes according to the chart which follows:

Component	Reaction 1 (Control)	Reaction 2	Reaction 3
Buffer	6 ml	3 ml	3 ml
Pellet in tube 3 (resuspended in buffer)	—	3 ml	—
Supernatant from tube 4	—	—	3ml
Methylene Blue	2-3 drops	2-3 drops	2-3 drops

Mix each of the tubes well, then add 1 ml of vegetable or mineral oil to the top of each. Incubate the tubes in a 37°C water bath several hours to overnight. Compare the tubes and record your observations.

Results

The experiment is generally very reliable. Use of pea roots rather than shoots avoids possible problems with chloroplast contamination.

Although different portions of pea seedlings are used in modules 6 and 8 of this unit, don't be tempted to save unused tissue from one experiment for the other experiment unless it can be used immediately. Fresh tissue gives the best results. Seedlings should be about seven days old (in any case, not older than twelve days).

Yield of mitochondria is difficult to predict and depends on a number of factors including the effectiveness of homogenization. For this reason, two dilutions of the mitochondrial fraction are assayed for marker enzyme activity.

12

PRINCIPLE OF CHROMATOGRAPHY

One of the most generally useful methods of protein fractionation involves chromatography, a technique that was originally developed for the fractionation of low molecualr weight components, such as sugars and amino acids. The most common type of chromatography used to separate small molecules is known as partition chromatography. In the most convenient form of this technique, a drop of sample is applied as a spot to a sheet of absorbent paper (in *paper chromatography*) or to a sheet of plastic or glass that has been covered with a thin layer of inert absorbent material such as cellulose or silica gel (*thin-layer chromatography*). Then a mixture of solvents, such as water and an alcohol, is allowed to permeate the sheet from one edge. As the solvents move across the sheet, they pick up those molecules in the sample that are soluble in them. The solvents are selected so that one of them absorbs more strongly to the absorbent material than the other. Cosequently, molecules that are most soluble in the absorbed solvent are relatively retarded, while those that are most soluble in the other solvent move more quickly. To detect the location of the different molecules, the chromatograph is dried and then stained. Various forms of paper and thin-layer chromatography are still widely used to analyze many different small molecules.

Proteins are most often fractionated by *column chromatography*, in which a mixture of proteins in solution is passed through a column containing a porous solid matrix and the different proteins are retarded to different extents by their interaction with the matrix. In this way, different proteins can be separately collected as they flow out of the bottom of the column. Many types of matrices have been developed

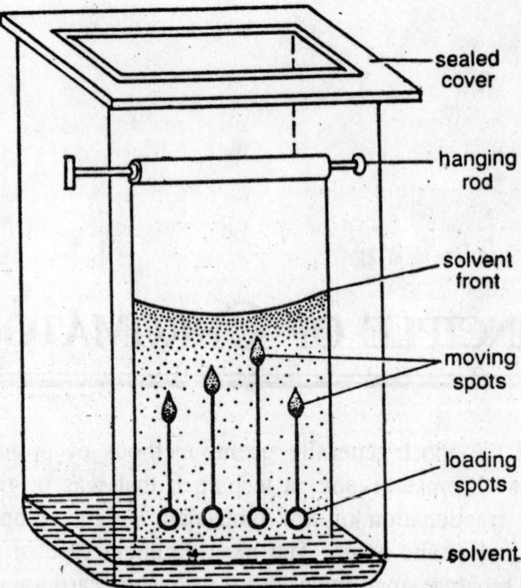

Fig. 12.1. Paper chromatography.

that allow proteins to be separated without altering their native structure. These matrices discriminate between different proteins by change size, or their ability to bind to particular chemical groups on the matrix.

The separation of small molecules by *paper chromatography*. After the sample has been applied to the origin and dried, a solution containing a mixture of two solvents is allowed to flow slowly through the paper by capillary action. Different components in the sample move at different rates in the paper according to their relative solubility in the solvent that is preferentially absorbed by the paper.

The separation of molecues by *column chromatography*. The sample is appplied to the top of a cylindrical column made of glass or plastic containing a permeable solid matrix immersed in solvent. Then a large amount of solvent is pumped slowly through the column and is collected in separate tubes as it emerges from the bottom. Various components of the sample travel at different rates through the column and are thereby fractionated.

How Cells are Studied

Schematic drawing of some different types of chromatographic matrix. In ion-exchange chromatography (A), the insoluble matrix carries ionic charges that retard molecules of opposite charge. Matrices commonly used for separating proteins are diethylaminoethylcellulose (DEAEcellulose), which is positively charged; and carboxymethyl cellulose (CM-cellulose) and phospho cellulose, which are negatively charged. The strength of the association between the dissolved molecules and the ion-exchange matrix depends on both the ionic strength and the pH of the eluting solution, which may be therefore varied in a systematic fashion to achieve an effective separation. In gel filtration chromatography (B), the matrix is inert but porous. Molecules that are small enough to penetrate into the matrix have a larger volume of solvent available to them and therefore travel more slowly through the column. Beads of cross-linked polysaccharide (dextran or agarose) are available commerically in a wide range of pore sizes, making them suitable for the fractionation of molecules of various molecular weights, from less than 500 to over $5' \times 10^6$. Affinity chromatography (C) utilises an insoluble matrix that is covalently linked to specific ligand, such as an antibody molecule or an enzyme substrate, that will bind a specific protein. Enormous purifications are often achieved in a single pass through such an affinity column.

Many different types of matrices are commercially available for this purpose. Ion-exchange columns are packed with small beads that carry either a positive or negative charge, so that proteins are fractionated according to the arrangement of charges on their surface. Hydrophobic columns are packed with beads from which hydrophobic side chains protrude, so that proteins with exposed hydrophobic regions are retarded. Gel-filtration columns are packed with tiny porous beads and separate proteins according to their size: molecules that are small enough to enter the pores percolate inside successive beads as they travel, while large molecules remain between the beads and therefore flow more rapidly through the column, emerging first. In addition to providing a means of separating molecules, gel-filtration chromatography is a convenient way to determine their size.

A particular protein cannot usually be separated from a mixture in a single step by column chromatography because each step generally increases the proportion of the protein in the mixture by no more than 20-fold. Since most proteins represent less than 1/1000 of the total cellular protein, it is usually necessary to use several different types of column in succession in order to purify them. A far more efficient procedure, known as affinity chromatography, takes advantage of the biologically important binding interactions that occur on protein surfaces. For example, if an enzyme substrate is covalently coupled to an inert matrix, such as a polysaccharide bead, the enzyme can often be retained by the matrix along with very few other proteins. In a similar way, specific antibodies can be coupled to a matrix, which can then be used to purify molecules recognised by the antibodies. Because of the great specificity of such affinity columns, 1000- to 10,000-fold purifications can sometimes be achieved in a single pass through the column.

Typical results obtained when three different chromatographic steps are used in succession to purify a protein. In this example, a whole cell extract was first fractionated by allowing it to percolate through an ion-exchange resin packed into a column (A). The column was washed and the bound proteins were then eluted by passing a solution containing a gradually increasing concentration of salt onto the top of the column. Proteins with the lowest affinity for the ion-exchange resin passed directly through the column and were collected in the earliest wash fractions eluted from the botton of the column. The remaining proteins were eluted in sequence according to their affinity for the resin—those proteins binding the tightest to the resin requiring the highest concentration of salt to remove them. The protein of interest eluted in a narrow peak and was detected by its enzymatic activity. The fractions with activity were pooled and then applied to a second, gel-filtration column (B). The elution position of the still impure protein was again determined by its enzymatic activity and the active fractions pooled and purified to homogeneity on an affinity column (C) that contained an immobilized substrate of the enzyme.

Terminology

In order to clarify the discussion of chromatography a knowledge of the terms likely to be encountered is helpful. The term *adsorbent* is often applied to the stationary phase. Although this usage is only correct when it refers to a solid stationary phase having the property of adsorption, the term adsorbent is sometimes used even when reffering to a liquid stationary phase or to a mixture of a liquid and solid. The word *sorbent* can be used more properly to include any stationary phase, whether it is liquid or solid. The term support is correctly used to indicate the solid scaffolding which holds a liquid stationary phase.

The mobile phase which causes the *solute* or *sample* to migrate through the stationary phase is called the *solvent* or *developer*. The forward edge of the solvent is called the *solvent front*. The point where the sample is placed at the start of the operation is the *origin*.

The position of a solute on a chromatogram is usually indicated by its R_f value which is the ratio of the distance it has migrated from the origin to that of the solvent front. Thus,

$$R_f = \frac{d(\text{solute})}{d(\text{solvent})}$$

where d(solute) is the distance travelled by the solute and the d(solvent) is the distance which the solvent front has moved from the origin. Under specific conditions R_f values are a characteristic of each chemical entity. Occasionally, in special cases, one may encounter another type of value such as R_g. This term may be used when a known solute is employed as a marker or standard and the positions of the other solutes are compared to it rather than to the solvent front. The solute on a chromatogram cannot usually be seen visually unless it happens to be a highly coloured substance. Thus, the solute must often be *detected* or *visualized* by some other means.

To collect the solutes that have been separated on a chromatogram, the solvent can be allowed to flow through the entire stationary phase until the desired solutes are carried out also. This process is called *elution* and the eluted mobile phase is the *effluent*.

Theory

The technique of chromatography is based on the fact that the molecules of solute in a sample are in a state of equilibrium between the solvent and the stationary phase. In a given stationary phase and solvent system, the distribution of the solute between the two phases is a property of the solute and may differ more or less from one compound to another. As a solvent or mobile phase is passed through a stationary phase, those solutes having a greater affinity for the stationary phase will be more retarded in their flow than those having less affinity. Thus, a separation of different solutes will occur which may be complete at the end of the operation.

Although the discussion of chromatographic theory which follows refers primarily to adsorption and partition processes, many aspects can be applied to ion-exchange and gel-chromatography. However, because of the importance of the latter two processes in experimental biochemistry, they will be discussed in some detail by themselves.

Distribution of Solute

The distribution of a solute between the two phases of a chromatogram is referred to as the *distribution coefficient* and is usually indicated by the letter K. By convention,

$$K = \frac{\text{conc. solute in mobile phase}}{\text{conc. solute in stationary phase}}$$

A value of 0.2 for K means that the stationary phase contains five times as much solute as the mobile phase.

The distribution of solute between the two phases can also be expressed in terms of an *effective distribution coefficient*, B. This term refers to the total amount of solute found in the mobile phase divided by the total amount found in the stationary phase. It can be expressed as

$$B = K\frac{\text{vol. mobile phase}}{\text{vol. stationary phase}}$$

The operation of a chromatogram can be described in terms of a large number of equilibrations. Each of these equilibrations can be thought of as occupying a layer or portion of the chromatogram, be it a column or a thin layer, and each layer is referred to as a *theoretical plate*. The length of each of these portions or layers in a direction perpendicular to the direction of solvent flow is called the height equivalent to a theoretical plate, or HETP. The latter term is a measure of the efficiency of a chromatogram.

For purposes of explanation consider an ideal chromatogram existing as a column of a solid stationary phase and a liquid mobile phase. The column is 7 cm length and contains 1 ml of solvent per centimeter. If 1 ml of solvent containing 128μg of a solute is added to the top of the column, Step 1, 1 ml of solute will be eluted from the bottom of the column and 7 ml of solvent will remain in the column. If the effective distribution coefficient B is equal to 1, then the solute will be equally distributed between the stationary and mobile phases.Thus, 64μg of solute will be found in each of the two phases of the top 1 cm of the column. If an additional 1 ml of solvent, without solute, is then added to the top of the column, another 1 ml of solute will flow out of the column. One will find, as seen in Step 2, that the 64μg of solute that had been in the mobile phase of the top layer has moved into 1-cm layer second from the top. Thus, 64 μg of sloute will be found in each of the top two 1-cm layers and will be equally distributed between adsorbent and solvent in each of those layers. The addition of another 1 ml of solvent to the column as illustrated in Step 3 will carry the 32 μg of solute contained in the solvent of the top layer into the second layer while the moving solvent will carry the 32 μg of solute from the solvent phase of the second layer into the third layer.

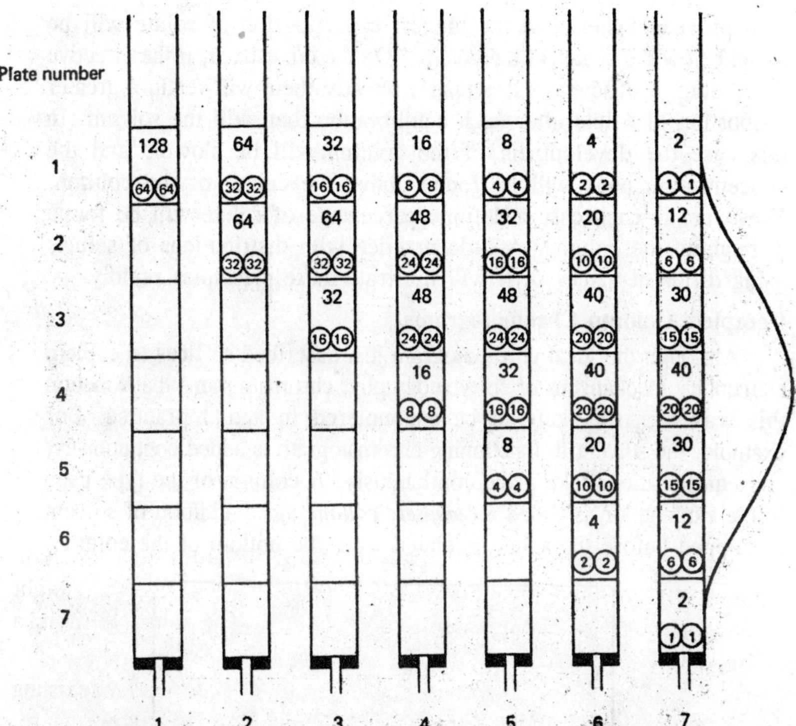

Fig. 12.2. Hypothetical distribution of solute during chromatography in a complete column.

The 32 μg of solute left in the top layer will now distribute itself equally between the two phases as will the 32 μg of solute carried into the third layer. In the second layer one will find the 32μg of solute carried from the solvent phase 64μg of the top layer and the 32 μg that had been left behind from the stationary phase. Thus, one will still find 64 μg of solute in the second layer where it is in a state of equilibrium between the two phases. The distribution of solute following the addition of each 1 ml of solvent can be observed in each of the additional steps. It can be seen in Step 7 that the solute is distributed throughout the column and that the highest concentration is in the center of the column. From the curve drawn alongside the column of Step 7 one can seen that the concentrations approximate a Gaussian distribution.

If the effective distribution coefficient in the above example is greater than 1, a greater fraction of solute will pass through with the solvent after each equilibration. The development of the column will then proceed faster, and the highest concentration of solute will be found below the center of the column. On the other hand, if the effective distribution coefficient is less than1, the adsorbent will retain a greater proportion of solute after each equilibration than will the solvent. In this case the development of the column will be slower, and the concentration peak will be found above the center of the column. However, in each case a definite percentage of solute will be found throughout the column. A little practice with distributions of solutes using different values of B will illustrate these principles rapidly.

Complete Column Chromatograms

Although the ideal chromatogram just described suffices to explain elementary mechanisms of chromatography, chromatograms that function this way are not likely to be encountered in actual practice. For example, the solvent in functioning chromatogram is added continuously, and equilibrations take place continuously. A column of the type used in the explanation is called a *complete column* since addition of solvent is stopped before the solute is eluted from the bottom of the column.

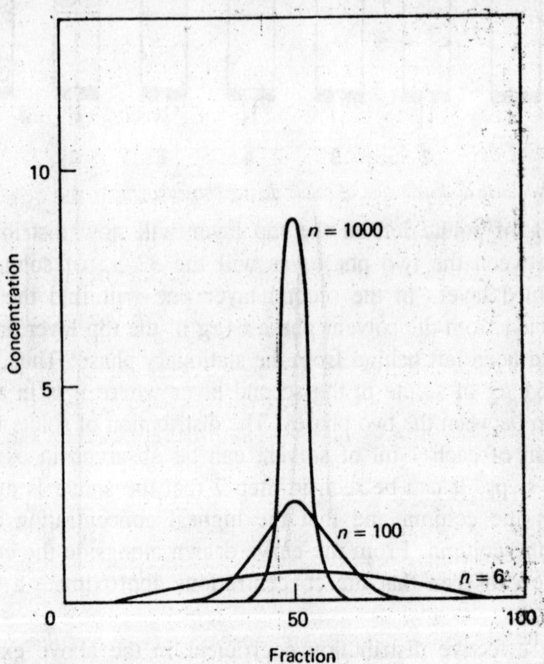

Fig. 12.3. Distribution of solute in a column chromatogram as a function of the number of equilibrations. The letter, n, refers to the number of equilibrations or theoretical plates.

Recovery of fractions of solute from such a column can be made by extruding the column from the tube and dividing it into sections, a process not likely to be chosen for preparative purposes. A somewhat analogous situation occurs in the case of thin-layer or -paper chromatograms where the development is terminated before the solutes reach the end of the support. The type of chromatogram where solutes are eluted is called an *elution chromatogram*; there are differences in the behaviour of these two types of chromatograms.

In the case of the complete column just described, the sample spreads over a greater portion of the column as the number of equilibrations, or theoretical plates, increases; however, the *degree* of spreading decreases as this happens. The same phenomenon can be expressed mathematically as the amount of solute per unit of the chromatogram by equation

$$C = \frac{\sqrt{n(B+1)}}{L\sqrt{2\pi B}}$$

where C is the fraction of total solute per unit length of column, n is the number of equilibrations or theoretical plates, B is the effective distribution coefficient, and L is the total length of the column. The equation can also be expressed as

$$C = \sqrt{\frac{n}{2\pi p(L-p)}}$$

where p is the position of the peak of the solute. These equations show that the degree of spreading decreases as the function of the square root of the number of equilibrations. In an actual chromatogram, equilibrations are taking place continuously and at least several hundred equilibrating layers of plates are present.

Resolving Power

One other concept in chromatography of value to the biochemist is that of *resolving power*. Essentially resolving power indicates the ability of a chromatogram to separate solutes that differ only slightly in their effective distribution coefficient.

The resolving power of a chromatogram is related to the values of the effective distribution coefficients of the compounds to be separated. The difference in these values may be used as a fair criterion of the resolving power of a given chromatogram. Relative resolving power of a complete chromatogram may be expressed in mathematical terms as

$$\text{Relative resolving power} = \frac{\sqrt{Bn}}{B+1} = \sqrt{nR_f(1-R_f)}$$

One may see from the above equation that the resolving power is proportional to the square root of the number of theoretical plates contained in the chromatogram; therefore lengthening a chromatogram would increase its relative resolving power.

Adsorption and Partition Processes

The distribution of solute between the mobile phase and the stationary phase can involve either adsorption or partition processes depending upon the type of stationary phase employed. In the partition process the solute is in equilibrium between two liquids—the liquid solvent and the liquid of the stationary phase—and its distribution is determined by its solubility in each. Those solutes which are more soluble in the solvent will migrate faster than those that are more soluble in the stationary phase. The partition system is more analogous to the case of the distribution of the solute between two mutually exclusive solvents in a separatory funnel and, in fact, the partition coefficient of a solute can be measured in a separartory funnel. Those compounds which differ in their partition coefficient should be separable. Since the partition coefficient for a compound is a constant for a given liquid—liquid pair, its relative distribution does not change even when its concentration does. In addition, its rate of movement is independent of concentration.

In adsorption the solute is in equilibrium between a solid adsorbent and a liquid solvent, and the behaviour is more complicated. Adsorption involves a variety of forces. In the case of neutral solutes, the principal force is dipole-dipole attraction with hydrogen bonding being involved in some instances. Where acidic or basic solutes are involved, ionic forces are chiefly responsible for the interaction. The solute then is caught between the attractive forces of the adsorbent on the one hand and of the solvent on the other. In addition, the solvent tends to complete with the solute for sites on the adsorbent; this results in a displacement effect. The vast interplay of forces competing for the solute often results in a variation in the distribution of solute between the two phases as a function of concentration. This behaviour is expressed in Figure where convex and concave curve represent the type of distribution encountered with adsorption while the straight line is that associated with partition. The convex curve is more commonly found in adsorption chromatograms and results in a phenomenon called "tailing" where parts of the solute lag behind the major portion.

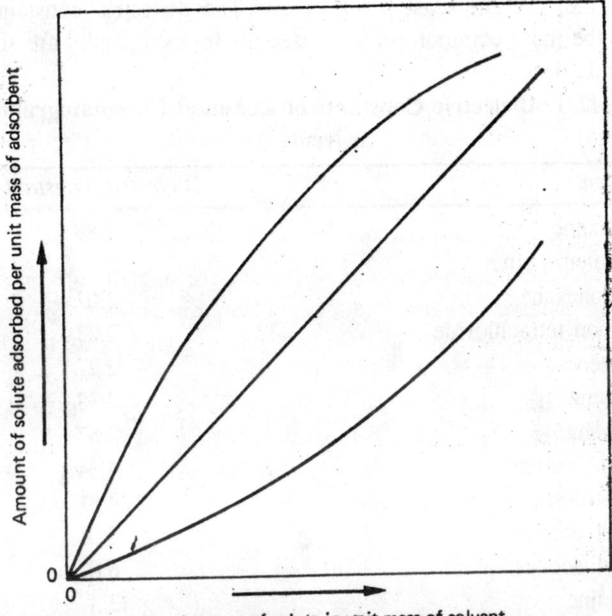

Fig. 12.4. Distribution of solute between the adsorbent and the solvent as a function of solute concentration showing the three most common types of distribution.

In order to choose between adsorption and partition chromatography, the experimenter must consider the objective of the operation and the compounds being separated. Adsorption processes are better for preparative work because larger quantities can be manipulated. Partition processes are dependent on the solubility of the solute in two liquid; therefore partition processes are most effective in separating compounds which differ in their solubility properties. Thus, one would expect members of a homologous series to be separated best by partition chromatography. Adsorption chromatography, on the other hand, is more effective in separating compounds that differ mainly in their configuration or charge.

The effectiveness of most solvents in separating solutes in adsorption or partition chromatography is related to their dielectric constants. These constants are dependent upon the degree of polarity of the compound but also are affected by hydrogen bonding and polarization. Water, for example, is a very polar solvent because it has a very strong dipole and it processes hydrogen bonding. Alcohols and esters have weaker dipoles and less hydrogen bonding, and are, therefore, less polar, whereas the hydrocarbons, such as petroleum, ether and cyclohexane, are the least polar of all. The dielectric constants of some of the more common solvents used in chromatography are shown in Table 12.1.

Table 12.1. Dielectric Constants of Common Chromatographic solvents

Solvent	*Dielectric constant*
n-Hexane	1.89
Petroleum ether	ca 2
Cyclohexane	2.02
Carbon tetrachloride	2.23
Benzene	2.27
Toulene	2.38
o-Xylene	2.57
Diethyl ether	4.34
Chloroform	4.81
Butyl acetate	5.01
Ethyl acetate	6.02
Pyridine	12.3
n-Butanol	17.1
n-Propanol	20.1
Acetone	20.7
Ethanol	24.3
Methanol	32.6
Water	78.5

A good rule-of-thumb with respect to the polarity of organic solvents is that polarity increases as the number of functional groups increases and decreases as the molecular weight increases. In addition, various salts have a high degree of polarity because of their positive and negative charges. Lastly, it should be pointed out that the elutroprism (the tendency to elute solutes) of a solvent system may be altered by combining solvents of different dielectric constants.

Column Chromatography

Column chromatography has proven to be one of the most useful techniques in biochemistry for the separation and purification of a wide range of compounds. In column chromatography a thin layer of sample is placed on a stationary phase held in a tube of suitable material, usually glass. The chromatogram is developed by permitting a solvent to flow through the stationary phase. The solutes separate as

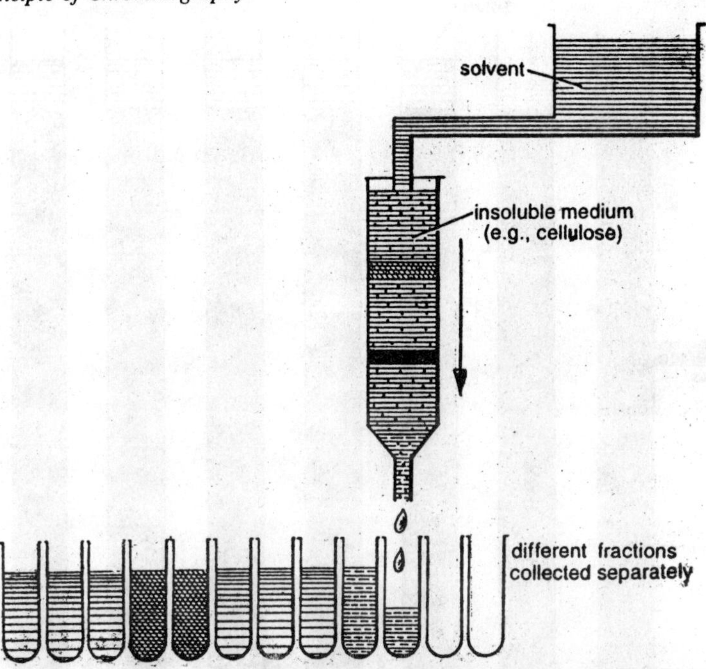

Fig. 12.5. Column chromatography.

bands in the chromatogram and are usually collected as separate fractions as they emerge from the column. The fastest moving component would be collected first; the slowest; last. This elution or flowing type of chromatogram is of more practical; value in experimental biochemistry than the complete type that was described during the discussion of chromatographic theory.

Discussion of the elution column chromatogram necessitates definition of two additional terms. These are the retention or hold up volume and the eluting volume. *Retention* or *holdup volume*, usually indicated by *v*, is equal to the volume of the mobile phase or solvent present in the column at any given time; *v* remains constant for any given column. *Eluting volume*, usually indicated by V, refers to volume of solvent that must be added to a column to produce the peak of a solute in the effluent. Of course, V is the function of the effective distribution coefficient or the R_f value of the compound being chromatographed. The relationship of these terms can be expressed mathematically as

$$\frac{V}{v} = \frac{B+1}{B}$$

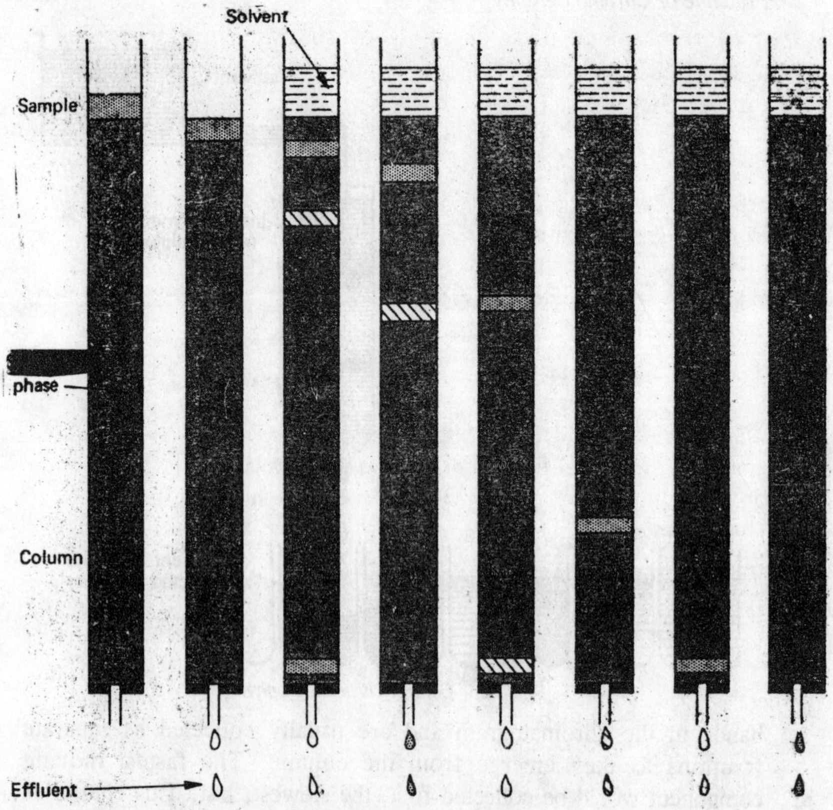

Fig. 12.6. A column chromatogram illustrating the separation and elution of three solutes of different R_f values.

and $$\frac{V}{v} = \frac{1}{R_f}$$

The term V/v is equal to the number of holdup volumes necessary to elute a given solute. Although it is the function of the effective distribution coefficient, V/v does not depend on the length of the column. The holdup volume can be determined experimentally by measuring the volume of solvent required to elute a dye that has no affinity for the stationary phase. Thus, it can be seen from these relationships that the volume of solvent required to elute a given solute can be calculated from the reciprocal of its R_f value of the holdup volume of the column is known.

As in the case of the complete column, the resolving power of a flowing column is a function of the square root of the number of theoretical plates in the column. However, the equation for its calculation differs slightly from that for the complete column and is

$$\sqrt{\frac{n}{B+1}}$$

where n is the number of theoretical plates and B is the effective distribution coeffficient of the compound being chromatographed. The resolving power of a flowing column may also be calculated from equation:

$$\sqrt{\frac{n(V-v)}{V}}$$

By substituting numerical values for the terms in the appropriate equations it can be seen that the maximum resolving power of an elution column is about twice that of a complete column.

The number of theoretical plates in an elution column can be calculated from the expression

$$n = \frac{8V(V-v)}{w^2}$$

where *w* is equal to the volume contained in the band of solute measured at 36.8% of the total height of the band in an elution diagram. The other terms are identical to those defined earlier.

Through the use of the above equations it is possible to calculate the volume of solvent required to elute a given solute and the volume of effluent which contains the solute. Furthermore, it is possible to determine the effect of increasing or decreasing the length of a column.

Choice of Chromatographic System

The first step in any chromatographic procedure is to choose a system that will effectively separate the substances in the mixture. If the compounds are known, the choice of system can frequently be made by consulting appropriate methods in the literature. If the components of the mixture are not known, a choice must be made on the basis of the predictable components. Where a mixture contains a component or components of interest as a small fraction of the total mixture, it is often desirable to remove the bulk of the extraneous substances by methods other than chromatography. The resulting partially purified mixture can then be purified further by appropriate chromatographic techniques.

As mentioned earlier, adsorption techniques are usually better for preparative work than are partition techniques. In addition, ion exchange is often a very powerful technique for preparative work. Ion exchange is frequently used in the purification of proteins from cell or tissue extracts and will be discussed in some detail later.

Columns

Columns are usually constructed of glass and may be designed in various lengths and diameters. Some are extremely simple; others are made more sophisticated, with the use of pressure, solvent reservoirs, fritted glass or nylon plates, water-jackets, and so on. In choosing a column for chromatography, the ratio of the length to the diameter of

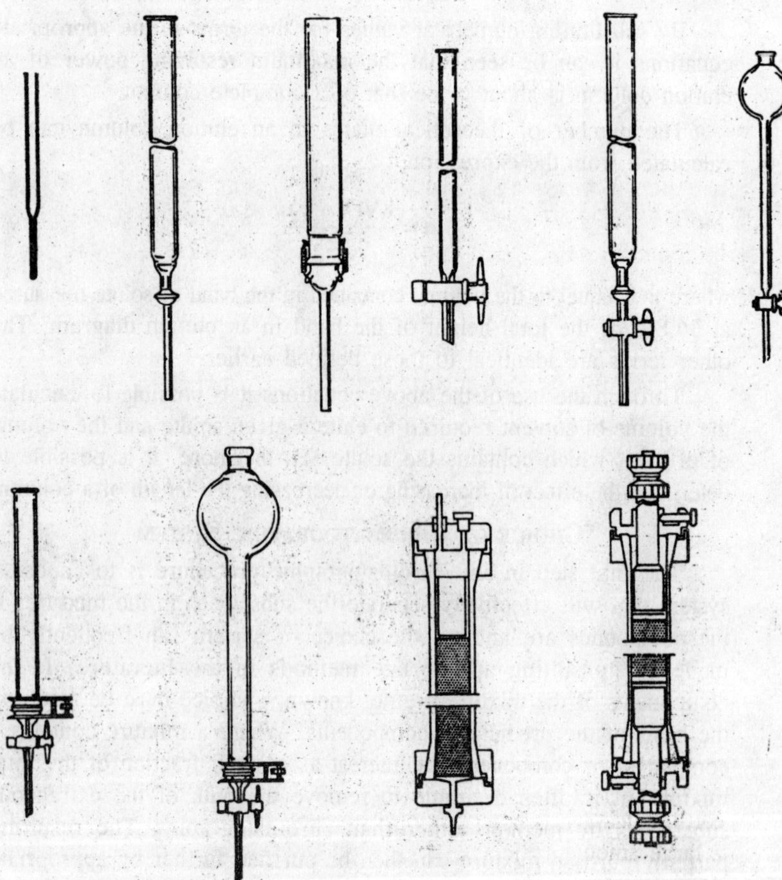

Fig. 12.7. Diagrams of various types of columns used for chromatography.

the column is sometimes important. An acceptable ratio depends on the ease of separation of the solutes. Shorter columns are easier to pack, but satisfactory separations often require longer columns whose ratio of length to diameter is 100—200 to 1.

Stationary Phase

The stationary phase of a column may be composed of either true adsorbents, partition agents, ion-exchange resins or cellulose, or gels. The theoretical aspects of adsorbents and partition agents were discussed earlier. Ion-exchange agents and dextran or polyacrylamide gels differ in many details from the traditional chromatographic adsorbents.

The more commonly encountered stationary phases are listed in Table 12.2. Of the true adsorbents, alumina and silica gel have been used most. The capacities and properties of these adsorbents can be controlled or "activated" by modifying their surface-water content.The degree of activation is indicated by a scale (called the Brockman index) from I to V where I is the most active and V the least active. For polar adsorbents, the most active contains the least water. The particle size of these adosrbents may vary from about 50 to 200 mesh. More detailed information on the treatment of specific adsorbents can be found in most of the general references on chromatography listed at the end of the chapter.

Table 12.2. Stationary Phases used in Adsorption and partition column chromatography.

Adsorbents
Neutral alumina
Basic alumina
Acidic alumina
Hydroxylapatite (calcium phosphate)
Magnesium silicate
Silica gel (silicic acid)
Partition agents
Diatomaceous earth (Celite, Filter-cel, super-cel)
Silica gel (silicic acid)
Cellulose

In the case of partition systems the support is usually diatomaceous earth, silica gel, or cellulose. The support should be capable of holding a large amount of the polar or nonpolar stationary phase and should not interact with it.

Solvents

Solvents should be of high purity. For columns containing an adsorbent as the stationary phase, the solvent can be a single compound or an appropriate mixture of compounds.

In the case of partition chromatography it is necessary to equilibrate the solvent system with the liquid used as the stationary phase. The usual procedure is to mix thoroughly the pair of immiscible liquids in a separatory funnel and then separate the two phases. The support for the column is usually mixed with the more polar phase while the second or less polar phase is used as the solvent system. It is important here that the equilibration of the two liquids be done at the same temperature as that at which the column is to be operated since a change in temperature will affect the mutual solubility of the two phases.

The solvent system must be chosen on the basis of the solubility of the sample, since the solutes must distribute themselves between the two liquids. An ideal value for the ratio of the solubility in the mobile phase to that in the stationary phase is 1 : 100. This ideal cannot always be realized, however, and a lower ratio must often be used.

Operation of Chromatographic System

Setting up Columns

The first step in setting up a column is to place a plug in the bottom of the tube. With simple columns a wad of glass wool or cotton will suffice. Other columns are made with a plate of fritted glass or with provision for a nylon or Teflon filter disc. Any tendency for particles of the stationary phase to clog the pores of the plug or disc may be reduced by placing a layer of fine glass beads, washed sand, or a filter-paper discover the plug.

Columns containing an adsorbent as the stationary phase may be packed either wet or dry. In the latter method the dry adsorbent is packed into the column a little at a time using a glass rod, wooden peg, or stopper held on the end of a rod as a tamper. Following the packing of the adsorbent, it is washed with the solvent to be used as the mobile phase.

Columns are more commonly packed wet by introducing the adsorbent in a slurry of the solvent. As the adsorbent settles, the excess solvent is eluted from the column and additional slurry is added until the column of adsorbent is built up to the desired length. An alternate method is to add dry adsorbent directly to the tube which already contains the solvent and to let the adsorbent settle out as described in the former procedure. In any case it is important to maintain a continuous column.

Columns containing partition agents are more difficult to prepare than those with adsorbents. In the case of supports composed of silica gel or diatomaceous earth, the stationary phase is usually mixed directly by adding the equilibrated stationary phase to the support in a ratio between 0.5 and 1.0 by weight (stationary phase support). Best mixing can be accomplished by grinding with a mortar and pestle or using a suitable blender. A slurry of this product and the mobile phase is prepared and poured into the column where it can be packed by gravity or by tamping with a suitable plunger. Usually columns of silica gel are packed by gravity whereas those of diatomaceous earth require tamping.

With cellulose, columns are prepared by adding a slurry of cellulose in acetone to the glass tube. The column is packed by gravity and excess acetone drained off the equilibrated solvent.

Once the column has been packed, the stationary phase is protected from disturbance by covering the top surface with a disc of filter paper, glass wool, small glass beads, or clean sand.

Introduction of Sample

The sample to be chromatographed should be added to the top of the column in a minimum amount of solvent and allowed to seep slowly into the stationary phase. A pipet or Pasteur pipet can be used to introduce the sample. After the sample has percolated into the stationary phase, a small volume of solvent is added in order to take the sample below the surface. This prevents dilution of the sample into a large volume of solvent.

Development

The column is developed by adding a large volume of solvent to the top of the column and allowing it to flow at a convenient rate. The rate is determined by the characteristics of the particular system being used and by the size of the columns. Better patterns of elution are obtained with slow rates of flow than with rapid rates. Tailing of bands in partition systems usuallly indicates a lack of equilibration that can be remedied by decreasing the flow rate. Where tailing occurs in adsorption systems, it may be corrected by increasing the polarity of the solvent system in a stepwise fashion. A broad change in the polarity sometimes results in shifting the order in which the solutes migrate and must be avoided.

One convinient method for increasing the rate of development of a column is to employ a gradual change in the composition of the solvent system. This method, called *gradient elution*, may involve a change in the ratio of two solvents or an increase in the concentration of a particular solvent such as a salt. Various pieces of apparatus have been designed to provide suitable solvent gradients.

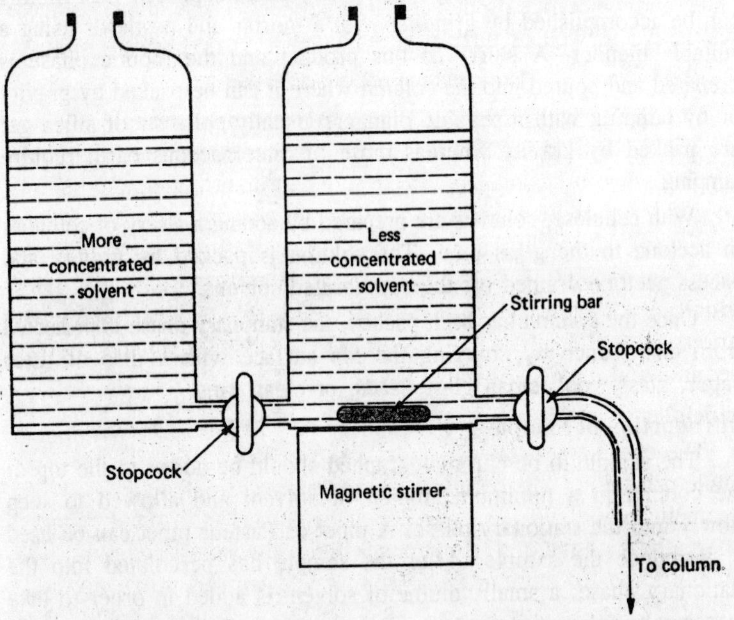

Fig. 12.8. Diagram of a gradient maker for column chromatographic development.

Collection of Effluent

The simplest method of collecting the solutes eluted from a chromatographic column is to introduce suitable reservoirs manually, such as test tubes, flasks, or beakers, to catch the effluent. The effluent can be divided into fraction on the basis of volume, drops, weight, or time. Although this method of collection is not a problem if the column is short and its flow rate is relatively fast, it becomes inconvenient with slow flow rates, particularly if the column must be operated in a cold room. Because of these inconveniences mechanical fraction collectors have been designed.

Mechanical fraction collectors are available from a number of manufacturers in models of varying sophistication. Usually fractions can be collected at preset time intervals, by volume, or by the number of drops. A typical fraction collector is pictured.

Detection

Solutes eluted from columns in different fractions can be detected by either direct or indirect methods, depending upon the nature of the solute and objective of the chromtographer. These methods have quantitative as well as qualitative application.

LIQUID CHROMATOGRAPHY

Liquid chromatography (LC) is a chromatographic technique which utilizes a liquid mobile phase to separate mixtures of compounds. Recent advances in pump technology have increased liquid chromatography's utility in the clinical laboratory. The separations that were once performed on long columns requiring hours of elution time are now feasible in minutes with the use of high-pressure pumps and small particle-size columns. This chapter discusses the basic principles, the various types, the important parameters, and the common equipment used in liquid chromatography.

Principles

The chromatography of a components of a mixture is based on simple principles. Separation of the components of a mixture may be accomplished because each component interacts with its environment differently from the other compounds under the same conditions. The solubility and the miscibility of compounds are the primary factors affecting their interactions.

The polarity of individual molecules determines their solubility and miscibility. The simplest organic compounds, alkanes and alkenes, are nonpolar. A good example of a nonpolar compound is hexane.

```
  H H H H H H
  | | | | | |
H—C—C—C—C—C—C—H
  | | | | | |
  H H H H H H
```

Factors affecting polarity are the presence or absence of electron donating or withdrawing groups, conjugation, and molecular symmetry. Examples of electron withdrawing groups and molecular symmetry affecting polarity and hexanol (left, below) and hexanoic acid (right, below), the latter being the more polar.

```
    H  H  H  H  H  H          H  H  H  H  H  H
    |  |  |  |  |  |          |  |  |  |  |  |
H—C—C—C—C—C—C—OH   H—C—C—C—C—C—C—OH
    |  |  |  |  |  |          |  |  |  |  |  |
    H  H  H  H  H  H          H  H  H  H  H  H
```

Most biological molecules vary from slightly polar to polar. Water is a very polar molecule because of its molecular structure which results in a permanent dipole.

The polarity of molecules is predictive of their general solubility : the rule "like dissolves like" may be applied in liquid chromatography for selection of appropriate mobile and stationary phases.

The components of a liquid chromatography column may be itemized as follows:

1. Stationary phase—may be a solid or a liquid.
2. Mobile phase—is a liquid.
3. Support—is most often a solid.

Separation of solutes may be achieved if the stationary and mobile phases are selected so that separation of the solutes occurs due to either differences in solubility of the solutes between the stationary and mobile phases or differences in adsorption on the stationary phase.

As a discrete band of sample-containing solution is forced through the column, the portion of solution high in concentration of a specific component is equilibrated with the stationary phase and the mobile phase; the next increment of mobile phase containing less of this component will extract a portion of the adsorbed component into the mobile phase. If every specific component enters into this "back and forth" exchange to different degrees and at different speeds from all other components, separation is achieved.

Types of Liquid Chromatography

There are four basic types of liquid chromatogrphy; liquid-solid (adsorption), liquid-liquid (partition), ion-exchange, and gel permeation (molecular sieving).

Adsorption

Adsorption chromatography is performed with a liquid mobile phase and a solid stationary phase which reversibly adsorbs solutes. Common examples of stationary phases are silica gel, porous glass beads, and alumina. The mobile phase is usually a relatively nonpolar solvent mixture. This technique is applicable when sample components vary widely in polarity, e.g., lipids.

Partition

Partition chromatography is performed with a liquid-coated stationary phase which is immiscible with the mobile phase. The relative distribution of the sample components between the mobile phase and the stationary phase determines the relative separation. Usually, stationary liquid phases and polar while the mobile phase is nonpolar (normal phase). Water and polyethylene glycol are typical normal phase stationary phases. Hexane and chloroform are common normal-phase mobile phases. It is possible to damage a column by removing the polar coating if a polar mobile phase is used. A very nonpolar molecule would move rapidly through a normal-phase column if the mobile phase were hexane. The nonpolar molecule would have a greater attraction for the nonpolar hexane than for the polar groups of the stationary phase.

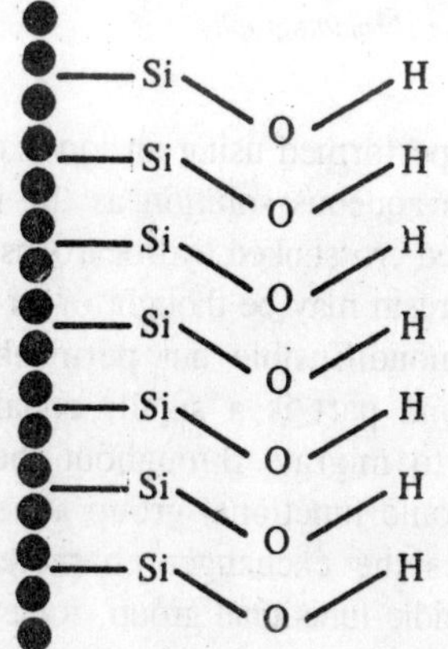

Fig. 12.9. Normal phase liquid chromatography.

Another form of partition chromatography, reverse phase, consists of a nonpolar stationary phase and a polar mobile phase. Typical mobile phases in this form of chromatography are water, acetonitrile, and methanol. Common stationary phases are hydrocarbon chains chemically bonded to the support. Partition chromatography is employed when sample components have known differences in solubility, e.g., drugs. Figure 12.9 illustrates the molecular structures of typical stationary and mobile phases employed in reverse-phase partition chromatography.

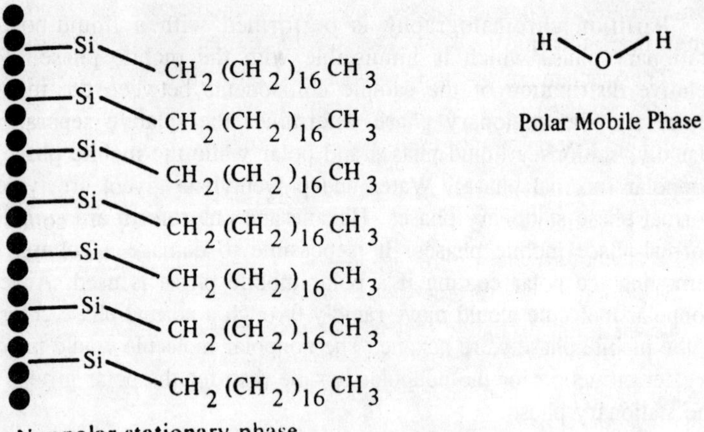

—Si—$CH_2(CH_2)_{16}CH_3$
—Si—$CH_2(CH_2)_{16}CH_3$
—Si—$CH_2(CH_2)_{16}CH_3$
—Si—$CH_2(CH_2)_{16}CH_3$
—Si—$CH_2(CH_2)_{16}CH_3$
—Si—$CH_2(CH_2)_{16}CH_3$
—Si—$CH_2(CH_2)_{16}CH_3$

H—O—H

Polar Mobile Phase

Nonpolar stationary phase

Fig. 12.10. Reverse phase chromatography.

Ion-Exchange

Ion-exchange chromatography is performed using an ion-exchange resins as the stationary phase and an aqueous solution as the mobile phase. The resins are highly polymerized crosslinked hydrocarbons which contain ionized functional group. The resin may be thought of as having two distinct parts; one is a large, nondiffusible but permeable ion (basic resin structure), and the second part is a small, equally but oppositely charged ion that is free to migrate throughout the resin structure during the exchange. The ionic functional group attached to the structure of the resin determines the exchanger characteristics. Cation resins are produced when an acidic functional group, for example

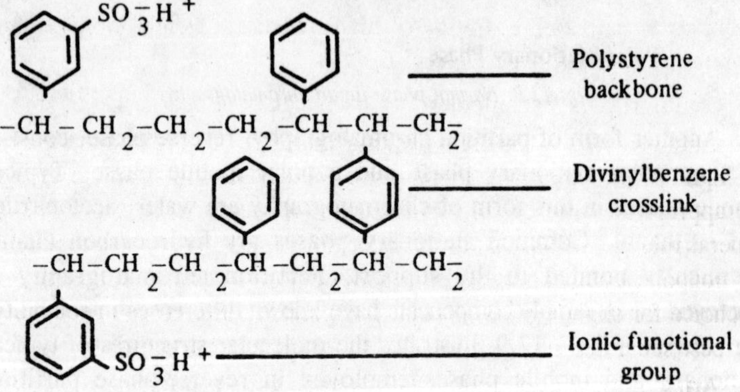

Fig. 12.11. Cationic ion exchange resin.

($-SO_3H$), is attached to the resin structure. The acidic hydrogen is available for exchange. Conversely, anion exchangers are produced when basic functional groups are attached to the resin structure.

The mechanism of the exchange is illustrated as follows :

$X^- + R^+ Y^-$ $\quad$ $Y^- + R^+X^-$ $\quad$ (anion exchange)

$X^+ + R^-Y^+$ $\quad$ $Y^+ + R^-X^+$ $\quad$ (cation exchange)

where

X = sample ion

Y = mobile-phase ion

R = ionic sites on exchanger

In an anion exchange system, the sample ion X^- is in competition with the mobile phase ion, Y-, for the ionic sites, R+, on the ionic exchanger. In cationic-exchange chromatography, the sample cations, X+ are competing with the mobile-phase ions, Y+, for the ionic sites. R-, on the ion exchanger. Solutes interacting weakly with the resin in the presence of the mobilephase ions are eluted rapidly, while solutes that react strongly with the resin are retained much longer, pH and ionic strength of the mobile phase determine the solute ion-resin interaction. Clinically, this technique is used for separation of small ions, small hormones, heme-breakdown products, and amino acids.

Gel Permeation

Gel permeation is a *mechanical* sorting of molecules based on molecular size. Size separation is accomplished with a porous packing gel through which the smallest components in the sample migrate into the smallest pores of the gel while the large molecules are encluded from the gel and elute from the column early in the separation. This technique is used for separation of high molecular weight enzymes and proteins from other serum constituents.

Chromatographic Parameters

Columns vary with their packing material, according to the type of liquid chromatography to be performed. It is important to choose the appropriate type of column for the compounds to be separated. A general rule to follow is : adsorption chromatography is most applicable for nonpolar compounds, while partition chromatography is the method of choice for slightly polar to polar compounds. Any ionized compounds are best separated utilizing ion-exchange chromatography.

An additional parameter which is useful for controlling separation is temperature. Increasing temperature may in some instances improve component separation and will reduce solvent viscosity which results in reduced column back pressure.

There are two basic methods for solvent delivery. The first is isocratic solvent delivery in which a mobile phase of fixed composition is delivered to the column at a set rate throughout the chromatogram. The second method is gradient solvent delivery in which one component of the mobile phase is varied in concentration throughout the chromatogram.

Equipment for Liquid Chromatography

The solvent delivery system may be one of the three basic types. The first is a syringe pump which is driven by a worm-screw drive. This mechanism yields a precise, pulseless flow. Another type is the single-piston reciprocating pump which maintains constant pressure by a pressure-sensitive feedback circuit. The third type of delivery system is a dual-reciprocating pump which has two pistons diametrically positioned and driven by the same cam. Both the single piston and the

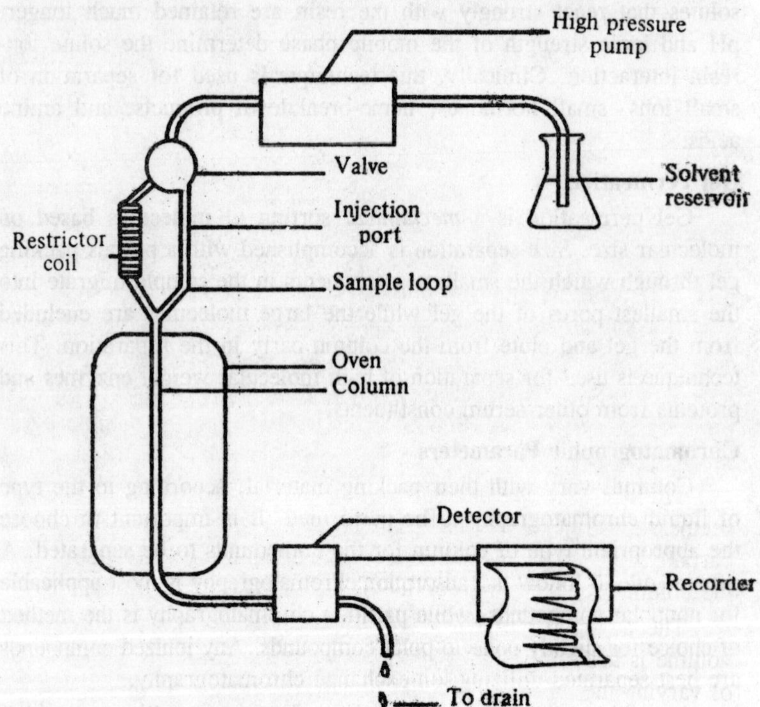

Fig. 12.12. Basic liquid chromatograph components.

dual-piston pump have a pulsed flow which is controlled by various damping mechanisms, resulting in low baseline noise. Table 12.3 illustrates the advantages and disadvantages of each solvent delivery system.

Table 12.3. Solvent Delivery Systems.

Solvent Delivery System	*Advantages*	*Disadvantages*
Syringe pump	Even solvent flow rate Low flow rates precisely controlled	Unable to replenish solvent without shutting system down: high pressure not attainable.
Reciprocating pump	Solvent reservoir easily refilled during operation; high pressures attainable; relative ease of operation	Pulsing solvent flow
Dual-headed reciprocating pump	Same as the reciprocating-pump system	Small amount of pulsing in the flow of the solvent

There are three basic types of sample injector systems for liquid chromatographs. The first is the on-column injection system which consists of a siliconized rubber disk, and a septum through which the sample is loaded into the flowing mobile phase at a point just above the column bed using a micro high-pressure syringe. Upkeep of this system is difficult because of leaks and punctured septums which cannot be replaced unless the pumping mechanism is shut down. Small particles of septum material, tend to collect at the head of the column, causing high back pressure and peak broadening.

The second type of injector is a fixed-loop injector in which the sample is loaded into the injector with a syringe to completely fill the injector sample loop. When the valve is turned to the "inject" position, the total volume of the sample loop is washed onto the column by the mobile phase.

The third type is the syringe-loop injector in which the sample, premeasured by microsyringe, is loaded into the sample loop which is partially filled with mobile phase. The sample is then flushed on to the column when the injector is placed in the "inject" position.

The advantage of the fixed-loop injector is that a reproducible volume is applied to the column ; but this system lacks the flexibility of varying the injection volume. Both loop injector systems, however, allow ease of sample injection against high solvent pressure and are superior to the septum injector.

There are four basic detectors used for liquid chromatography; fixed wavelength adsorption, multiple wavelength adosorption, fluorescent, and electrochemical.

Table 12.4. Liquid chromatograph detectors.

Detector	*Advantages*	*Disadvantages*
Fluorometric	High sensitivity for fluorescent compounds	Moderately specific
Electrochemical	Extremely sensitive	Too sensitive
	Good specificity	Mobile phase must be an electrolyte
Fixed wavelength	Cheap	Poor specificity
	Stable	
	Relatively sensitive	
Multiple wavelength	Good specificity	Moderately sensitive

The fixed-wavelength UV-visible detector generally consists of a mercury lamp light source and utilizes filters to select the desired wavelength. The multiple-wavelength UV-visible detector utilizes a monochromator to obtain the wavelengths of choice. Both the fixed and multiple wavelength detectors are spectrophotometers utilizing a microflow cell as a cuvette. The fluorometric detector is a fluorometer adapted with a microflow cell to detect fluorescent compounds.

Electrochemical detectors are employed to detect compounds that can be oxidized or reduced, thus changing the conductivity of the compounds as compared to the solvent flowing through the flow cell.

Advantages and Disadvantages

The first advantage of liquid chromatography, as compared to other types of chromatography, is the simple sample preparation. A single extraction is usually sufficient, and derivatives usually are not required, thereby removing a step in which experimental error is likely to be introduced. This technique may be used at ambient or slightly above-ambient temperatures. Because of the relatively low temperatures involved in liquid chromatography, sample stability is of little concern. Finally the instrumentation for liquid chromatography is reliable, and routine maintenance is simple compared to gas chromatography.

Two major disadvantages of liquid chromatography are the lack of sensitivity to some compounds and the requirement of moderately expensive equipment to perform analyses.

Gel Chromatography

The application of various gels to chromatography has proven to be a valuable tool for the separation of solutes differing in molecular weight. By this technique larger molecules are eluted from a column of the gel before the smaller molecules. Thus, the method has wide use in the purification of proteins, in desalting protein-containing solutions, and in the determination of molecular weights.

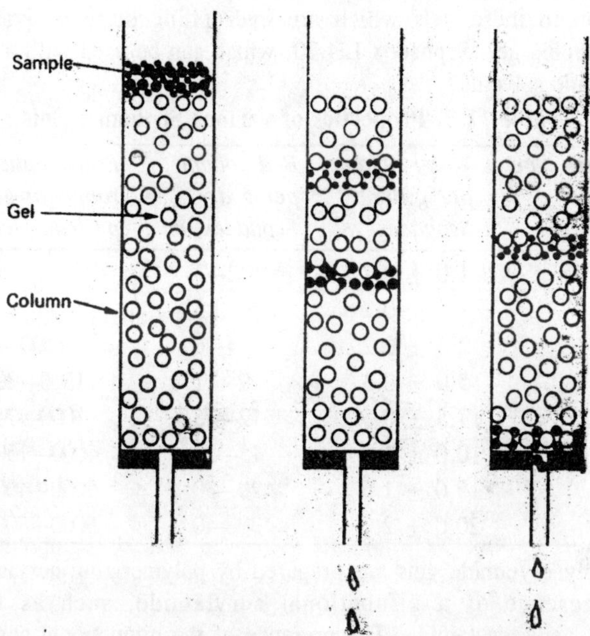

Fig. 12.13. Gel filtration column showing the distribution of solute particles of two different molecular weights during development.

The first theoretical approaches to gel chromatography considered the behaviour of solutes in a column of gel as being governed primarily by steric effects. Small molecules were viewed as being able to penetrate the regions between the gel chains whereas large molecules, because of their size, were prevented from entering. However, the steric approach does not take into consideration other factors which influence the distribution of many solutes between the gel and the mobile phase. These factors can be included in a unified approach to the theory of gel chromatography by interpreting the latter as a kind of partition chromatography.

Gels presently used in chromatography consist of three principal types (a) dextran, (b) polyacrylamide, and (c) agarose.

Dextran gels are prepared by reacting alkaline solutions of crystalline dextran with epichlorohydrin. The properties of the gel can be controlled by adjusting the concentration of epichlorohydrin and the molecular weight and concentration of dextran. Dextran gels are available commercially under the name Sephadex and are prepared in different particle sizes and with different degrees of crosslinkage. In addition to these gels which are hydrophilic there is available a hydrophobic gel, Sephadex LH-20, which can be used with a number of organic solvents.

Table 12.5. Properties of various Sephadex Gels.

Sephadex type	*Water-regain per g dry Sephadex, ml*	*Bed volume per g dry Sephadex, ml*	*Fractionation range for globular proteins, mol.wt.*
G-10	1.0 ± 0.1	2—3	—700
G-15	1.5 ± 0.2	2.5—3.5	—1500
G-25	2.5 ± 0.2	4—6	1000—5000
G-50	5.0 ± 0.3	9—11	1500—30,000
G-75	7.5 ± 0.5	12—15	3000—70,000
G-100	10.0 ± 1.0	15—20	4000—150,000
G-150	15.0 ± 1.5	20—30	5000—400,000
G-200	20.0 ± 2.0	30—40	5000—800,000

Polyacrylamide gels are prepared by polymerizing acrylamide in the presence of a bifunctional acrylamide, such as N, N′-methylenebisacrylamide. The presence of the bifunctional acrylamide results in crosslinking between different chains of the polymer, and a three-dimensional array of chains is built up. A carboxylic acid amide group on every other carbon atom introduces polarity into the gel. Polyacrylamide gels for chromatography are available commercially under the name Bio-gel. They are available with different degrees of crosslinking and, thus, different pore sizes.

Agarose is a linear polysaccharide of D-galactose and 3, 6-anhydro-1-galactose which is purified from agar, an extract of certain seaweeds. It is free of ionizable groups and can be prepared as fine beads. It is available commercially under the name Sepharose. Agarose may be used for the separation of solutes of greater molecular weight that can be separated with the dextran or polyacrylamide gels.

Table 12.6. Properties of Various Bio-gels.

Bio-gel type	*Water-regain per g dry Bio-gel, ml*	*Bed volume per g dry Bio-gel, ml*	*Fractionation range for globular proteins, mol. wt.*
P-2	1.6	3.8	200—2000
p-4	2.6	6.1	500—4000
p-6	3.2	7.4	1000—5000
P-10	5.1	12.0	5000—17,000
P-20	5.4	13.0	10,000—30,000
P-30	6.2	14.0	20,000—50,000
P-60	6.8	18.0	30,000—70,000
P-100	7.5	22.0	40,000—100,000
P-150	9.0	27.0	50,000—150,000
P-200	13.5	47.0	80,000—300,000
P-300	2.0	70.0	100,000—400,000

Preparation of Gels for Chromatography

Both dextran and polyacrylamide gels must be hydrated before they can be used in a column. The water-regain values listed in Tables show the minimum quantities of water necessary to hydrate the gels, although 5-10 minutes this amount should be used to prepare gels. Those gels with a small pore size can be hydrated in a few hours at room temperature; however, those with large pore sizes require up to 24 hours. The process of hydration or swelling can be accelerated by placing the slurry of the gel in a boiling water bath. Boiling also helps to eliminate air bubbles which may become trapped in the slurry and which will cause an uneven flow of the column.

Selection of Columns

Selection of a column for gel chromatography is governed by its use. For example, short columns can be used for desalting purposes, whereas long columns may be necessary for critical separations. Very narrow columns should be avoided because they lead to wall effects in which the solvent has a tendency to flow faster in a thin layer along the glass wall. Wall effects can be minimized by coating the interior surface of the column with a solution of 1% dichlorodimethylsilane in benzene. The solution is added to the column at 60°C and, after standing for a few minutes, is decanted. The benzene is then evaporated in a drying oven and the entire process repeated again. Such a double coating is generally satisfactory for several months. In choosing a column one should also avoid those with large chambers at the outlet and those with fritted glass or porous plastic discs; these have a tendency to become clogged with the fine beads of gel.

Packing Columns

Packing columns with gels of low pore size rarely causes any difficulties. Thus, gels of Sephadex G-10 to G-50 and of Bio-gel P-2 to P-10 can be packed without special precautions. Usually the column is partially filled with the buffer solution to be used and the slurry containing the gel is poured into the top of the column. As the bed begins to form, the outlet at the bottom of the column is opened. Additional slurry can be added until the desired height of gel is obtained. The gel is then rinsed with several column volumes of buffer at a flow rate similar to that used for elution of the sample. The gel must never be allowed to dry out.

Gels of high pore size present more problems in packing . Improperly packed columns will not perform satisfactorily if they

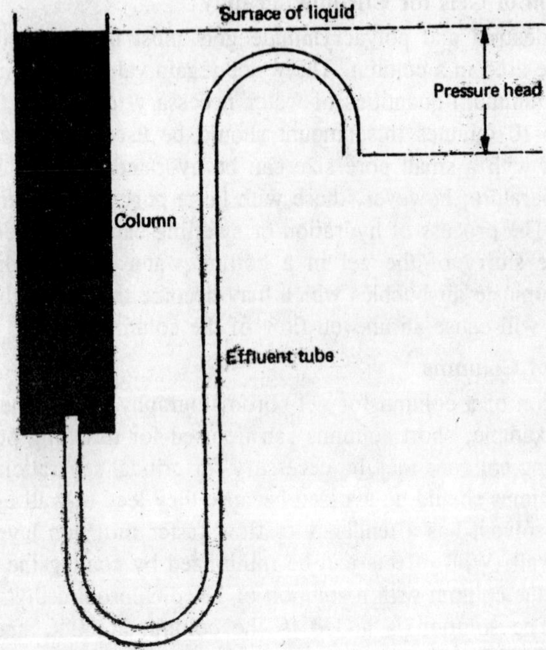

Fig. 12.14. Apparatus for packing a column with gel sorbent of high water-regain properties.

perform at all. A small amount of buffer is first poured into the column which is positioned vertically with the outlet tubing held above the top of the column. The slurry of swollen of gel is poured into the column, preferably down a glass rod. It is desirable to pour all of the slurry to produce a complete bed into the column at one time. An extension tube mounted on top of the column may be necessary to hold the excess slurry. The slurry in the column is stirred gently with a glass rod to remove any air bubbles. The outlet tube is next lowered below the surface of the liquid in the column and the buffer is allowed to flow slowly out. A slowly rising horizontal surface of the gel indicates uniform packing. The outlet tubing is then lowered gradually until the operating pressure to be used in the column is reached. This pressure should not exceed 100—150 mm in the case of the highest porosity gels but can be increased for gels of lower porosity. High pressure compresses the particles and impedes normal flow.

Repacking

Most columns containing gels can be reused without any special regeneration. However the column must be cleansed of unwanted solutes by sufficient washing with water or buffer.

After lengthy use some columns flow more slowly because of compression of the gel bed. Flow rates can be increased by backwashing or repacking the gel.

Sample Application and Elution

Several methods have been used for application of samples on gel beds. The simplest involves adjusting the buffer to the top of the bed level followed by the careful addition of the sample to the top of the bed using a suitable pipet. After allowing the sample to drain into the gel, a small amount of buffer is added to wash the sample completely into the gel. The buffer can then be added to the desired level, the column connected to a buffer reservoir, and the elution begun.

For analytical purposes the sample should be applied in as small a volume as possible. When the width of the zone of solute is not as critical, such as is the case in desalting, the size of the sample is less important.

Solutes that are uncharged can usually be eluted with distilled water, but those with charged groups usually require a developer with an ionic strength of 0.02 or more. The developer is necessary for best results because most gels contain a small number of charged groups which result in some adsorption of solutes.

For best results the rate of elution should be constant. This can be accomplished manually or by using a constant-pressure flask such as a Mariotte flask.

Elution of solutes from columns of gels can also be accomplished by reversed flow. In this method the sample is applied at the lower end of the chromatogram and solutes are washed upward through the gel and out the top of the column. This method is most useful where gels of the highest porosity are being used in order to increase their usually very slow flow rate.

Determination of Molecular Weights

The elution volume of globular proteins and other macromolecules is determined largely by their molecular weight and is a linear function of the logarithm of the molecular weight. In practice a suitable standard of known molecular weight is first applied to the column for calibration purposes. A standard as closely related as possible to the substance being studied should be chosen. Equations for the calculations of molecular weights of globular proteins have been derived for several different dextran gels. These are

Sephadex G-75: $\log_{10}$ mol. wt. = 5.624—0.752 (V/v)
Sephadex G-100: $\log_{10}$ mol. wt. = 5.941—0.847 (V/v)
Sephadex G-200: $\log_{10}$ mol. wt. = 6.698 —0.987 (V/v)

where V is equal to the elution volume and v is equal to the holdup volume. The holdup volume can be determined by measuring the volume of solvent required to chromatograph a coloured compound of very high molecular weight. Substances such as Blue Dextran 2000 (Pharmacia) with a molecular weight of 2.0×10^6 are available commercially for this purpose.

THIN-LAYER CHROMATOGRAPHY

Thin-layer chromatography has recently won widespread favour in biochemistry. It has been used extensively for qualitative, quantitative, and preparative purposes and can be applied successfully to the separation and identification of a wide range of organic compounds. TLC offers the advantages of quick separation, high sensitivity, simple equipment, and ready adaptability. TLC has replaced paper chromatography as the most useful analytical chromatographic method.

In TLC the stationary phase is attached to a suitable support such as glass or plastic film. The stationary phase can be a finely divided powder acting as an adsorbent or as a support for a liquid. Many powders are used as sorbents, but silica gel and cellulose are the most common. Silica gel can function as an adsorbent following heat "activation" and can be used quite successfully to separate many of the less polar compounds. Silica gel and cellulose, when properly coated with a liquid film, act as supports for partition chromatography of a wide range of polar and nonpolar compounds. After the sample is applied to the sorbent near one end of the thin-layer plate, the chromatogram is developed in an ascending fashion by immersing the lower margin of the plate in solvent inside an appropriate chamber. After the solvent has risen a suitable distance, the TLC plate is removed from the solvent, dried, and the separated spots visualized by any one of a variety of methods.

Operation of TLC System

Supporting Plates

Plates used for the support of chromatographic thin layers may be of glass, metal, or suitable plastic. Glass plates can be used in a number of sizes ranging from microscope slides to large plates. Microscope slides are the simplest and can be used to separate upto four components in a few minutes. More complex mixtures require larger plates. The latter are also more useful for preparative TLC where thicker layers of adsorbent must be applied. Flexible sheets of polyester film can also serve as bases for thin layers. Although they are more difficult to coat than glass plates, polyester film does not break if dropped.

Precoated plates of glass and film are now available commercially. Polyester film plates cannot be used with all system solvents, however.

Sorbents

In theory any stationary phase used in column chromatography can be used in TLC. In fact, TLC has been referred to as open-column chromatography. In practice true adsorbents and partition agents usually comprise the stationary phase or layer.

Adsorbents for TLC are much smaller in practice size than those employed for column chromatography. Silica gel is the most common adsorbent although basic alumina is of greater use for the separation of bases. Silica gel and cellulose are most commonly employed as partition agents although diatomaceous earth is sometimes used.

The layers of sorbents are usually held in place on the plates by various binders that are mixed in with the sorbent prior to layering. The most common binders is plaster of paris (mixed with sorbents at a level of 10 to 15%). Starch and certain organic polymers are also used as binders and result in much harder layers than those obtained with plaster of paris.

Preparation of Layers

Thin layers are usually prepared by layering a film of sorbent and water as a slurry on a clean plate and allowing it to dry. The thickness of the slurry should vary with the sorbent used and with the requirements of the chromatographic operation. Slurries that are too thick or too thin should be avoided. A consistency of pea soup is usually best. Most manufacturers recommend optimum water/sorbent ratios for individual products. The slurry is usually made by shaking the sorbent with water in a flask or by mixing in a blender.

Microscope slides can be coated simply by immersing the slide into the slurry, withdrawing it carefully, and allowing it to dry in a horizontal position. Larger plates can be layered by using commerically available apparatus or by employing the technique. In the latter method the tape is placed along the margins of the two opposite sides of the plate. As many layers of tape as necessary to obtain the desired thickness of sorbent are used. The slurry is poured out along one untaped edge of the plate and is spread out with a glass stirring rod held across the plate with both ends resting on the tape. After the slurry has been spread out evenly it can be allowed to dry for about 30 min. Following removal of the tapes the sorbent can be activated by heating at 110°C for 1 hour in an oven.

Prepared plates are available from commercial sources in a wide variety of sorbents and supporting plates. Layers are usually of excellent quality.

For partition chromatography thin layers can be impregnated with a number of different liquids. When water is chosen, the impregnation can be combined with drying the slurry by allowing it first to dry at room temperature followed by heating at 105°C for 10 min. Polar and nonpolar liquids can be impregnated into the layer by dipping the layers into the impregnating liquid after the layer has been dried to remove excess water. Usually a solution of the liquid in some volatile solvent is employed, such as 20% formamide in acetone. The impregnated layers are then removed carefully from the solution and are set in a hood to dry. No heating is used.

Substances called phosphors are sometimes included with the slurry of sorbent. These usually emit visible light when irradiated with ultraviolet light. Many solutes, particularly those that are aromatic or contain conjugated double bonds, will quench the emission of visible light. The presence of these solutes is indicated under ultraviolet irradiation by dark spots in a bright field.

Solvents

Solvents for TLC may be chosen on the basis of the guidelines discussed in the section on chromatographic theory and in the section on column chromatography. In partition systems the relative solubility of the solutes in the two liquid phases must be considered. Typical solvents systems for separating polar solutes include (a) isopropanol/ammonia/water (9 : 1 : 2); (b) butanol/acetic acid/water (4 : 1 : 5); and (c) phenol/water (4 : 1).

Solvents for TLC employing adsorbents may be chosen by running a preliminary series of tests with samples on microscope slides or small strips of film. The chromatogram is developed by different solvents in small scaled bottles. The best solvent systems may be further modified to yield even better results in a second series of tests. Then the best system can be chosen for more sophisticated separations. Optimum resolution is obtained near the center of the chromatogram.

Remember to consult the literature for help in choosing solvent systems.

Application of Samples

The first step in applying a sample to a TLC plate for one dimensional chromatography is to draw a very light pencil line with a straight edge 2 cm from one margin of the plate. Light pencil hash marks are next made at convenient intervals along the straight line (usually 1.5 to 3 cm between marks).

The samples should be applied at the hash marks with a fine capillary tube or a micropipet. Best results are obtained with 1- to 10-microliter (ml) samples and the spots on the layer should be as narrow in diameter as possible. The best procedure is to touch a very small drop to the sorbent and wait for it to dry before superimposing the next small drop. If the sample is applied in a volatile solvent, drying usually takes place very quickly. When less volatile solvents are used drying can be accelerated by blowing a stream of warm air across the plate from a hair dryer. In any event the TLC plate must be absolutely free of solvent before development is begun as its presence may adversely affect the separation.

For two-dimensional chromatography one sample is applied to one corner of the TLC plate 2 cm from each of the nearest two margins.

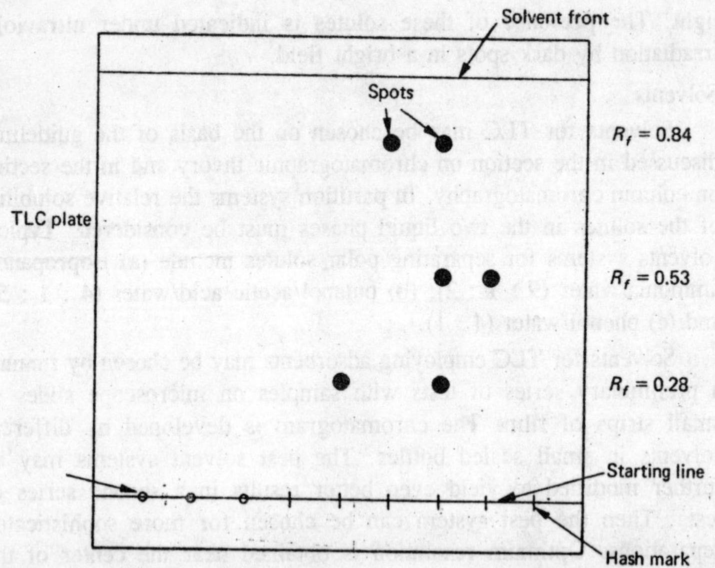

Fig. 12.15. A hypothetical thin-layer plate showing the origin, solvent front, and spots visualized after development.

Development

Thin-layer chromatograms can be developed in a wide variety of chambers. Microscope slide plates can be accommodated in a small jar saturated with the solvent. Larger TLC plates can be developed in chambers. The larger chambers require a period of time for the solvent to saturate the atmosphere of the chamber. This period may be shortened by lining the walls of the chamber with filter paper moistened with the solvent. Sandwich plates, which are frequently used as TLC chambers, have the advantage of not requiring pre-equilibration with solvent because their void space is so small. However, one must be sure that the edges of these are absolutely tight. Evaporation of solvent from around the edges adversely affect the flow of solvent along the margins of the TLC plate. The chamber should be placed in an area free of sharp temperature fluctuations.

When the chamber has been equilibrated, the solvent is adjusted to a depth of about 1cm and the TLC plate is placed in the chamber with the end nearest the samples standing level in the solvent. The cover of the chamber is set in place and development allowed to proceed. The progress of the chromatography can be observed by following the movement of the solvent front visually. Development

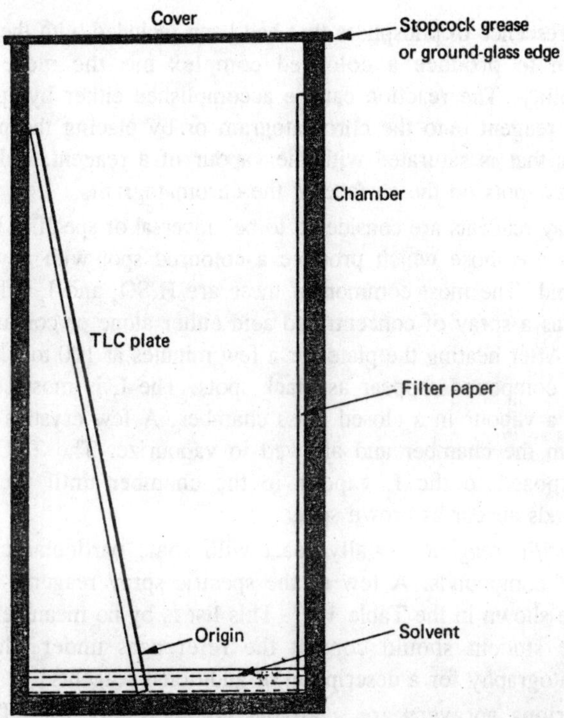

Fig. 12.16. Diagram of a thin-layer chromatography plate in a development chamber showing the location of the solvent and filter paper soaked with solvent to saturate the atmosphere of the chamber.

can be terminated when the solvent front is within about 2 cm of the upper margin of the plate. One must be certain that the level of solvent in the bottom of the chamber does not fall below the sorbent on the plate. On completion of the development the chromatogram is removed from the chamber, the solvent front is delineated with a soft pencil, and the plate is dried in a hood.

With two-dimensional chromatograms the TLC plate is developed in one direction by the first solvent and dried; then, after rotating the plate 90° in a clockwise direction, it is developed by the second solvent.

Detection

Chromophoric solutes of intense colour present no problems in visualization on TLC plates. Other solutes can be detected by their fluorescence under uv light or by their radioactivity. Most solutes, however, must be visualized by their reaction with some reagent to produce a coloured complex or by their quenching of light produced by the fluorescence of phosphors that had been included with the sorbent. Reaction to produce a coloured complex has the more general applicability. The reaction can be accomplished either by spraying a suitable reagent onto the chromatogram or by placing the plate in a chamber that is saturated with the vapour of a reagent. Solutes will appear as spots on the surface of the chromatogram.

Spray reagents are considered to be universal or specific. Universal reagents are those which produce a coloured spot with any organic compound. The most common of these are H_2SO_4 and I_2. The H_2SO_4 is used as a spray of concentrated acid either alone or containing 5% HNO_3. After heating the plate for a few minutes at 100 to 110°C, the organic compounds appear as black spots. The I_2 is most commonly used as a vapour in a closed glass chamber. A few crystals of I_2 are placed in the chamber and allowed to vapourize. The TLC plate is then exposed to the I_2 vapour in the chamber until the organic compounds appear as brown spots.

Specific reagents usually react with some particular classes or types of compounds. A few of the specific spray reagents and their uses are shown in the Table 12.7. This list is by no means exhaustive and the student should consult the references under Thin-Layer Chromatography for a description of additional reagents.

Various sprayers are available commercially for TLC spray reagents or they may be constructed in the laboratory. Most reference books on TLC contain instructions for their fabrication. In addition, pressurized cans of the more common spray reagents are available commercially.

To spray a TLC plate, first place it nearly vertically in a hood in front of some heavy wrapping paper or corrugated paper. The sprayer should be held 8 to 12 in. from the plate. Spraying should be done in sweeping motions across the plate. The plate must *never* be soaked since this will cause the spray to run on the plate and possibly move the solute.

When the solutes on a TLC plate have been visualized, the best procedure is to outline the spots with a soft pencil. This is absolutely necessary with those solutes detected using ultraviolet light and is advisable with others since some spots have a tendency to fade and disappear with time.

A record of the chromatogram can be made either by photography or by using a suitable photocopier. Colour slides or prints are recommended to preserve a record of coloured spots. **Otherwise**

Table 12.7. Specific Spray Reagents for Thin-layer Chromatography.

Reagent	*Preparartion*	*Procedure*	*Use*	*Colour*
Aniline phthalate	Dissolve 0.93 g aniline and 1.66g o-phthalic acid in 100 ml water saturated n-butanol	Spray soln. and heat for 10 min. at 105°C.	Reducing sugars	Various colours
Anisaldehyde in H_2SO_4	Mix 0.5 ml anisaldehyde and 0.5 ml conc. H_2SO_4 in 9 ml 95% ethanol	spray soln and heat for 10 min at 10–-100°C	Carbohydrates and steroids	Shades of blue
2,4-dinitrophenyl-hydrazine	Dissolve 0.1 g 2, 4,-di-nitrophenyl hydrazine in 100 ml 95% ethanol; add 1 ml conc. HCl	spray soln. and let dry at room temp.	Aldehydes and ketones	Yellow to red
Ninhydrin	Disolve 0.3 g ninhydrin in 100 ml n-butanol; add 3 ml glacial acetic acid	Spray soln. and heat for 10 min at 110°C	Amino acids, amino sugars, aminophos-phatides	Shades of violet except proline and hydroxy-proline, yellow.
Rhodamine B	Dissolve 0.5 Rhodamine B in 100 ml 95% ethanol	Spray soln. and dry at room tem.	Lipids	Dark violet spot on pink background in Uv*light
Antimony chloride in acetic acid	Mix equal weights of antimony trichloride and glacial acetic acid	Spray soln. and heat for 5 min. at 95°C	Steroids, lipids light.	Various colours in uv

uv*-ultraviolet,

photocopying is most convenient. If equipment for photorecording is not available, a tracing of the chromatogram may be made for the laboratory notebook.

The location of the spots on a TLC plate is defined by the R_f value. The center or density of the spot is used for purposes of measurement. This presents little difficulty with circular spots of symmetrical distribution. However, in the case of spots with no symmetry, such as those exhibiting tailing, a little common sense is necessary to locate the center of density.

The R_f values of some common amino acids in several different solvent systems are shown in Table 12.8.

Table 12.8. R_f Values of Common Amino Acids in Various Solvent Systems

	R_f			
Amino acid	*1*	*2*	*3*	*4*
Alanine	0.47	0.27	0.29	0.40
β-Alanine	0.33	0.27	0.30	0.29
Aspartic acid	0.55	0.21	0.06	0.06
Arginine HCl	0.04	0.08	0.19	0.07
Cysteic acid	0.69	0.14	0.04	0.21
Cystine	0.39	0.16	0.12	0.22
Glutamic acid	0.63	0.27	0.10	0.15
Glycine	0.43	0.22	0.24	0.34
Hystidine HCl	0.33	0.06	0.32	0.42
Hydroxyproline	0.44	0.20	0.38	0.31
Isoleucine	0.60	0.46	0.49	0.58
Leucine	0.61	0.47	0.48	0.58
Lysine HCl	0.03	0.05	0.09	0.11
Methionine	0.59	0.40	0.49	0.60
Phenylalanine	0.63	0.49	0.55	0.60
Proline	0.35	0.19	0.50	0.30
Serine	0.48	0.22	0.20	0.31
Threonine	0.50	0.25	0.26	0.40
Tryptophan	0.65	0.56	0.63	0.58
Tryosine	0.65	0.47	0.47	0.56
Valine	0.55	0.35	0.40	0.51

Errors

Some of the errors encountered in column chromatography occur also in TLC. Overloading, tailing, and changes in temperature are three of the problems common to both techniques, and their solutions are the same. However, in TLC, in addition to overloading of particular solutes, a physical overloading may also occur. This usually happens when the solutes to be separated constitute a small percentage of the total sample and the bulk of the contaminants are practically insoluble in the solvent system. The effect is to encase the solutes in the contaminant, from which they are only slowly extracted by the solvents ; the solutes, therefore, will be spread out over a long portion of chromatogram. When such results are obtained, the sample must be purified before chromatography is attempted.

Another error occassionally encountered in TLC, particularly in partition systems, results from a failure to control the amount of water in the liquid stationary phase of the TLC plate. The source of the error rests in the use of solvent systems that are saturated with water. Chromatograms operated under these conditions never have good reproducibility. The best solution to the problem is to use solvent systems which are unsaturated with respect to water.

Lastly, one must be sure to attain equilibrium of the atmosphere in the chromatographic chamber before beginning development. Failure to do so results in a serious lack of reproducibility.

SPECIAL TECHNIQUES

Preparative TLC

TLC can be used as a preparative technique, and quantities of solute up to 100g have been separated successfully under optimum conditions. Usually, however, quantities less than 2 g are chromatographed.

In preparative TLC the sample to be chromatographed is applied in a streak by drawing a capillary or a pipet containing the sample along a thin line 2.5 cm from the lower edge of the plate. The sample is developed in a direction perpendicular to the streak. The solutes must be visualized by some nondestructive means, following which the adsorbent containing the solute is scrapped off the glass plate. The solute is then eluted from the adsorbent.

Good sorbent layers for preparative TLC are difficult to prepare since they must be thicker than 1 mm. Unless the use of a large quantity of these TLC plates is anticipated, it seems advisable to purchase commercially prepared thick layers. These products generally are of high quality.

The spotting of the sample for preparative TLC is the most critical step, since volumes upto 2 ml must be applied without disturbing the layer. Many devices have been developed to accomplish this task. They are described in the literature and in commercial catalogs.

Development of plates for preparative TLC is similar to that for normal TLC except that the best solvent system should be ascertained by preliminary TLC on thin-layers. Following development, the chromatograms are dried in a hood with blowing air or with a current of nitrogen.

Visualization of preparative TLC plates is best accomplished by means of ultraviolet light on phosphor-containing layers or by exposure to vapours of I_2. In addition solutes may be detected by spraying a suitable reagent along one edge of the chromatogram (and sacrificing that portion) or by pressing a tape across the suspected bands and then spraying the tape with an appropriate reagent. The tape can then be used as a key to locate the bands on the plate.

Bands of solute can be scraped off the glass plate along with the adsorbent onto aluminium foil or powder paper. The solute is then eluted from the adsorbent with a suitable solvent. This last step can best be carried out by placing the scrapings in a sintered glass funnel or in a cone of filter paper in a glass funnel and washing the scrapings several times with the solvent.

Quantitative TLC

TLC plates can be assayed quantitatively, but the best precision one can expect has an error of 5 to 10%. Application of samples and development require particular care and attention to detail. A preliminary development of the plates before spotting the samples is often recommended to purify the adsorbent.

Assay of solute quantities can be performed on the thin-layer or following elution. Assays in place include the measurement of spot size and the measurement of spot density. To measure density, a photoelectric densitometer is used to scan the spots of solute and measure their density. Appropriate standards are measured also and the unknowns compared to these. Sorbent layers of uniform thickness are absolutely necessary. Assay of spot size is simple and does not require any instrumentation, although its results are not usually as precise. It is based on the observation that the square root of the area of a spot is directly proportional to the logarithm of the weight of the solute as in equation:

$$\sqrt{A} = m \log_{10} W + c$$

where A is the area of the spot, W is the weight of the solute, and m and c are proportionality constants for the substance being measured as determined from a calibration with a pure sample of the compound.

In addition to the spot-size and density methods, the quantity of solute can be measured in place by measuring radioactivity or by using a dilution method. In the latter method a series of dilutions of a standard solution of the compound to be assayed are chromatographed and the limit of detection on TLC is determined. A series of dilutions of the unknown are chromatographed and the solution at which the solute is just detectable is determined. The quantity of solute in the unknown is then calculated from the limit of detection and the dilution factor.

Quantitative TLC can also be carried out by eluting the solute from the TLC plate with an appropriate solvent and then determining the concentration spectrophotometrically. The recommended procedure is to chromatograph the unknown with a series of standard concentrations, visualize the spots, delineate the spots into rectangles of identical size, scrape the adsorbent within each rectangle off the plate with a razor blade onto a sheet of aluminium foil, transfer the adsorbent containing the solute to a sintered glass funnel, and wash the adsorbent with an appropriate solvent into a calibrated container. After bringing the extracts to a known volume with solvent, their spectral adsorbance can be determined in a spectrophotometer. The value for the unknown is then calculated on the basis of the adsorbances of the standards and appropriate blanks.

Gas Chromatography

Gas chromatography is a form of chromatography which employs an inert gas as the mobile phase and, generally, a liquid as the stationary phase. The general principles when using the liquid stationary phase are the same as those which apply to partition chromatography; in this form the technique is often referred to as gas-liquid chromatography (GLC).

Gas chromatography may be used to separate any compounds that can be vapourized without decomposition. This restriction is the limiting factor in chuosing gas chromatography as an analytical tool, although nonvolatile compounds can sometimes be converted into highly volatile derivatives. Gas chromatography has two advantages over other forms of chromatography : separations can be made in a much shorter period of time, and the technique can be made relatively precise quantitatively.

In GLC a column (either metal or glass) is placed with some inert packing which is coated with a layer of some nonvolatile hydrocarbon such as paraffin oil. The sample is introduced with the inert gas used as the mobile phase and is heated to volatilize its components. The vapourised components of the sample will dissolve in the liquid stationary phase but, because they are volatile, will equilibrate with the gaseous mobile phase, each with a definite

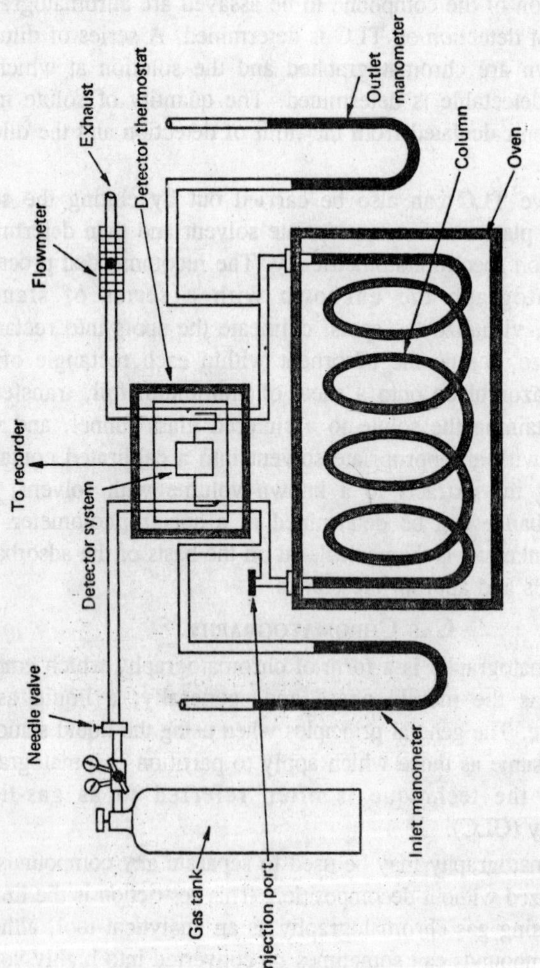

Fig. 12.17. Diagram of a simplified gas chromatography apparatus.

distribution coefficient. Thus, the conditions for a chromatographic separation are satisfied and the compounds will pass through the column at a rate determined by their distribution coefficients. Upon reaching the end of the column the compounds are detected, usually by means of thermal conductivity with a detector connected to a suitable strip-chart recorder. Thermal conductivity is convenient because most organic vapours have much lower thermal conductivities than carrier gases such as helium.

As mentioned earlier samples can be developed by gas chromatography much faster than those methods that employ a liquid mobile phase. The rapidity of development is related to the differences between diffusion rates of compounds in a gas and in a liquid. Diffusion rates are approximately 10,000 times greater in air than in water. Since the rate of development of a chromatogram is limited by the rate of the diffusion equilibrium between the stationary and mobile phases, the higher diffusion rates of the gas allows gas chromatograms to be developed much faster than those with a liquid mobile phase. Development times of gas chromatograms are as much as 1000 times faster than those with liquids.

Column and Stationary Phases

The fact that gas chromatography must be performed at elevated and controlled temperatures requires that the chromatographic column be placed in an oven. Thus, the column must often be coiled so that it can be accommodated even though a straight column would be preferable. Columns may be constructed of plastic, aluminium, copper, glass, or stainless steel. Stainless steel is most often used although glass is required for certain separations.

Columns are of two types; packed and open-tubular. Packed columns, as their name implies, are packed with a solid support for the liquid stationary phase. *Open tubular columns* are usually prepared by coating the wall of the column with the liquid stationary phase. The internal diameters of columns vary greatly according to their use. Preparative columns may be as large as 4 in, whereas some open-tubular columns are as narrow as 0.01 in.

The length of a column is governed by two main factors : the degree of separation desired and the pressure drop from the inlet end of the column to the outlet end. Generally, the longer to column the better will be the separation. The pressure drop, however, is also greater as the length of the column increases although it is also affected by the particle size of the support. When the pressure drop is large, the column is more difficult to operate. Thus, a compromise in column length must occasionally be made. Most packed columns are 4 to 15 ft in length.

Stationary Phases

Diatomaceous earth is usually used as support for packing a GLC column. Other materials such as crushed Teflon and fine glass beads have also been employed successfully. In open-tubular columns the surface of the tubing provides the support for the liquid stationary phase. Such columns are much longer than packed columns in order to obtain the necessary resolution of solutes.

Stationary phases must be chosen on the basis of the nature of the sample to be separated. Most of the references on gas chromatography, as well as commercial catalogs of GLC supplies, provide information to help make a proper choice. For compounds of biochemical interest silicone grease, paraffin oil, or polyethyleneglycols (Carbowaxes) are useful.

Preparation of Columns

Several methods are available for applying the stationary phase to the support. Each involves first, dissolving an amount of the stationary phase in a volatile solvent ; second, mixing the solution with the solid support; and third, evaporating the volatile solvent. The stationary phase usually comprises 5 to 20% by weight of the solid support.

The column is packed by slowly filling a suitable piece of tubing with the coated support after plugging the other end with glass wool. The tubing must be tapped often during the filling to pack the column evenly. When the column has been packed, the open end is plugged and the tubing is coiled to fit inside the oven.

Open-tubular columns are prepared by using an inert gas to force a solution of the stationary phase in some volatile solvent through the tubing. As the solution is forced through some of the solvent evaporates, leaving a film of the stationary phase on the wall of the column. Such a procedure requires the use of a specific amount of solution of known concentration.

All new columns must be *conditioned* before use. This is necessiated by the fact that new columns will bleed some stationary phase which will foul the detector. Conditioning is done by running carrier gas through the column, with the detector disconnected, at a temperature about 20°C higher than that at which the column will be operated. Conditioning normally takes about 10 to 12 hours.

Mobile Phase and Operation

Carrier Gas

The gas which is used as the mobile phase must be inert with respect to the components of the chromatographic system as well as to the components of the sample. Helium is the most commonly used carrier although chromatographs that have flame-ionization or electron-capture detectors require nitrogen or argon.

Temperature

The temperature of operation is one of the most important aspects of a gas chromatographic column. Since the R_f values of compounds rise with temperature, it becomes a useful variable. The selection of temperature is largely a matter of experimentation, although the preliminary choice can be made by starting with a temperature a few degrees below the boiling point of the presumed major component of the sample. A suitable compromise of temperature must often be made since columns at high temperatures operate more efficiently but separate less effectively. A satisfactory alternative can be made by *temperature programming*, the GLC equivalent of gradient elution. In temperature programming a predetermined pattern of raising the temperature during the development is established. The net result is to separate readily the more volatile solutes at the lower temperatures and then later develop the less volatile solutes at higher temperatures.

Detectors

Detectors for gas chromatography are of two types—integral and differential. The integral type indicates a sum of a property related to the mass of each compound developed by the column. The *differential type* indicates the instantaneous concentrations and is the type almost always encountered in commercial instruments. The detectors described below are of the latter type. The two most common detectors are the thermal-conductivity cell and the flame-ionization detector. The sensitivity of these detectors is about 2 mg and 10^{-5} mg, respectively.

The thermal conductivity cell consists of a double arrangement of fine platinum or alloy wires, maintained at a constant temperature, in the axis of a cylindrical channel. One channel is used as a reference channel with carrier gas only passed through it, whereas the other channel is the measuring channel used for the effluent of the column. The thermal conductivity cell is based on the fact that the temperature of the electrically heated wire depends not only on the electrical energy supplied but also on the thermal energy lost by radiation, conduction, and convection. The nature of the gas surrounding the wire will affect the loss of thermal energy by the wire. Thus, every variation in the composition of the gas changes the temperature of the wire and consequently its resistance. In practice, the column effluent containing the solute and the carrier gas has a lower conductivity than the carrier gas alone. Thus, the wire will heat when in contact with the solute, and the resistance of the wire will increase in comparision to that of the reference channel. The increase will result in a decrease in the flow of a current across the wire which then unbalances a Wheatstone bridge. The current required to balance the Wheatstone bridge is then measured by a recorder.

Another type of detector, with considerably better sensitivity than the thermal conductivity cell, is the flame-ionization detector. This detector employs a flame produced by burning hydrogen mixed with air. The effluent from the column is mixed into the hydrogen-air mixture before the gas is burned and will produce carbon dioxide as the result of the combustion of any organic compounds it may contain. The carbon dioxide molecules are then ionized and the ion concentration is measured by converting the changes in current into changes of potential. These changes can be highly amplified to produce a great increase in sensitivity. Since the response to the detector is linear, it is excellent for quantitative assays.

Operation

A constant flow of carrier gas is required for satisfactory operation of the gas chromatography column. Needle valves are available for this purpose and can be inserted into the line between the column and the diaphragm valve of the gas tank. The rate of flow can be monitored by using a bubble flowmeter on the exit port of the apparatus.

Selection of the best flow rate depends on the diameter of the column and to some extent on its individual idiosyncracies. Selection of a proper flow rate can be aided by chromatographing two closely boiling liquids such as benzene and carbon tetrachloride and adjusting the flow rate to produce good resolution. If the flow rate is too fast, resolution will be poor; if it is too slow, considerable tailing will result.

As discussed earlier the choice of a temperature for the operation of a gas chromatograph is largely a matter of experience. If the boiling point of the major component of the sample is known choose a temperature a few degrees lower. Many commercial gas chromatographs come equipped with a temperature programmer or can be modified with a temperature-programming accessory. This is a very desirable feature for a chromatograph. Usually the increase in temperature is linear and can be made at some chosen rate.

The sample is injected into the chromatograph through the injection port by means of a microsyringe. The injection port is covered by a silicone rubber septum which is punctured by the needle of the syringe. In order to vapourize the sample immediately the temperature of the injection port should be about 10°C higher than that of the column.

The amount of sample injected depends on the column and on the type of detector being used. With thermal-conductivity detectors the initial injection should be about 10 microliters (ml) of a 1% solution of the sample. When a flame-ionization detector is used then 1 ml of a similar solution should suffice. The injection of the sample should be done as quickly as possible since a slow injection will result in smearing the solutes through the column. On puncturing the rubber septum the plunger of the microsyringe must be held firmly to keep it from being pushed out by the gas pressure of the system.

Chromatograms are usually recorded by means of a stripchart recorder which is connected to an amplifier that increases the strength of the signal produced by the detector. Most recorders for gas chromatography have a 1 millivolt (mV) fullscale deflection. In most commercial instruments the signal may be controlled by an attenuator which can lower the signal in steps if it is too large to fit entirely on the chart paper. This is especially important where the highly sensitive flameionization detector is being used.

Identification of Components

Solutes in a sample assayed by gas chromatography may be identified by means of several different methods. The two most commonly used are the addition of the suspected compound to the sample and the use of retention times.

Retention time is analogous to the R_f value of other forms of chromatography. It is the time that is required for a compound to emerge from a column as compared to the time required by a standard air peak. An example of a recording of a chromatogram is illustrated in Figure. Although each compound should have the same retention time in a column at a given temperature and gas flow rate, in reality the time will vary somewhat. This difference is caused by the variability of other solutes in the sample which will affect the retention time of their neighbours and by the "age" of the column.

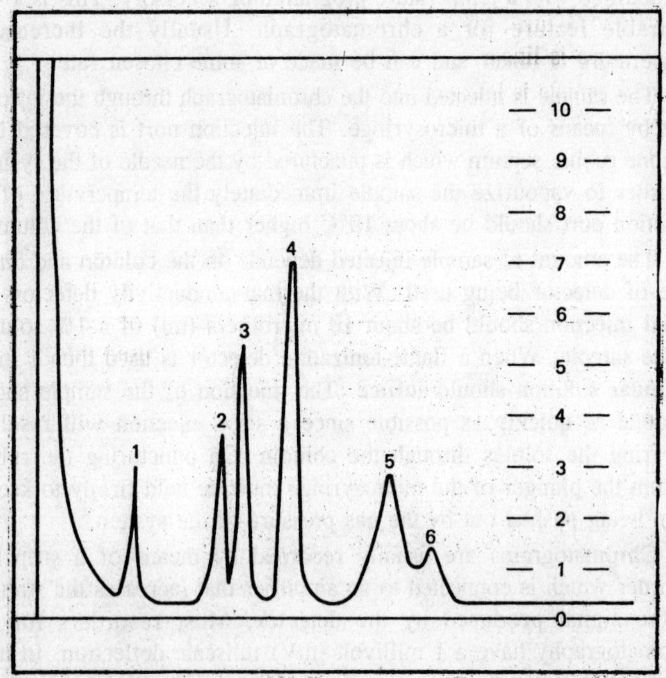

Fig. 12.18. Recorder trace of fatty acid methyl esters separated from a mixture by gas chromatography.

Compounds in a homologous series can be identified by plotting their retention times logarithmically as shown in Figure, where the values obtained for the methyl esters of fatty acids illustrated in Figure are plotted as a function of chain length.

Identification of a compound can be aided by adding a small amount of a known compound to the sample and chromatographing the mixture. If the unknown is the same as the known, then the height of the peak should increase and no additional peak or shoulder should be evident. If the unknown is different from the standard, then an additional peak or shoulder will appear on the recorder trace.

Quantitative Assays

The concentration of a compound separated by gas chromatography is usually determined by measuring the area of its peak. This area may be measured by (*a*) integration, (*b*) planimetry, (*c*) triangulation, and (*d*) cutting and weighing.

The most primitive and least precise of these is cutting and weighing which involves cutting out the peak from the recorder sheet with scissors

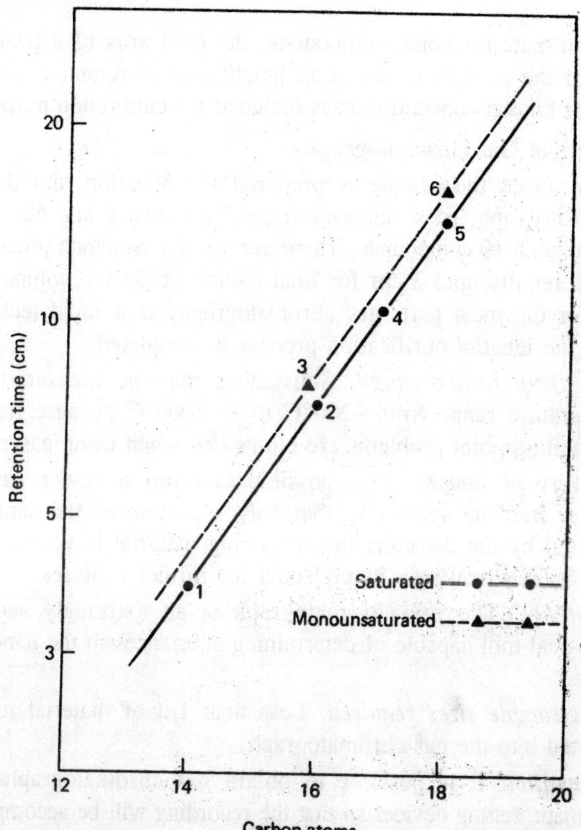

Fig. 12.19. A graph illustrating the relationship between the number of carbon atoms in homologues of fatty acids and the logrithm of their retention time in a gas chromatograph.

and weighing the paper on an analytical balance. The weight of the unknown can be calculated by comparision to the weight of a standard. This method is affected by the accuracy of the cutting and by variations in the thickness and moisture content of the chart paper.

Direct integration is the most precise method of measurement. This may be accomplished by mechanical or electronic means. The electronic method is the more precise (and the more expensive) of the two. These integrators are available from commercial sources and their operation is described by the manufacturers.

One fairly simple method of integration can be accomplished with a ruler, pencil, paper, and a calculator. This method is based on the observation that, for some compounds, the total area of a peak is a function of the product of the peak height and its retention time. A standard of known concentration is included for calibration purposes.

Advantages of Gas chromatography

1. *Speed.* Once the sample is prepared for injection into the gas chromatograph, most analyses require between 5 and 60 min to determine 8-10 compounds. There are a few uncommon procedures which require upto 24 hr for final elution of the last component ; but for the most part, gas chromatography is a rapid technique after the lengthy purification process is completed.
2. *Wide temperature range.* Substances may be measured in a temperature range from –200°C to + 1000°C because freezing and boiling-point problems are eliminated when using gases.
3. *Recovery of sample.* It is possible to insert a device called a splitter into the system so that only a portion of the sample is detected by the detector; the remaining material is vented to the outside, where it may be collected for further analyses.
4. *Sensitivity.* The gas chromatograph is an extremely sensitive analytical tool capable of determining substances in the nanogram range.
5. *Small sample sizes required.* Less than 1μl of material may be injected into the gas chromatograph.
6. *Automation.* It is possible to obtain gas chromatographs with automatic setting devices so that the recording will be accomplished automatically and the analyst need not be in attendance at the instrument for the entire run.

Disadvantages of Gas Chromatography

1. Lengthy preparation time for samples.
2. Contamination of detectors in the case of impure samples.
3. Time-consuming column preparation.

13

SPECTROPHOTOMETRY

Because of the similarity of function and use, both spectrophotometers and photoelectric colorimeters will be presented here. These instruments that are similar in design with respect to having many identical components can be markedly different in application and versatility. The colorimeter, which utilizes a filter as a monochromator, has proven to be a most rugged and stable instrument requiring very little or specialized usage because of the fixed wavelength of the monochromatic system. This wavelength can be changed only by changing the filter. The spectrophotometers, on the other hand, are more versatile but require much more maintenance. This is because the condition of the light source and the dispersing element is directly related to the quality of the monochromatic light; therefore this system requires closer monitoring and more maintenance. However, the dispersion element enables the selection of a continuously variable source of monochromatic light with either a simple manual or automatic adjustment.

It seems that as the efficiency of the monochromatic system to produce a narrow half-bandpass with a relative high percentage of light transmitted increases, the complexity of the associated optics and light source also increases. As the sensitivity of the detectors increases, the requirements for a more stable power supply and meter system also increase. Because of the interrelation of the aforementioned, it must understood that a good knowledge of the component parts, and their function as a part of that system, is a prerequisite to understanding the system. With a knowledge of these systems, principles of operations, and of the laws that govern the absorption of light, the student is more able to cope with the problems of photometry.

PRINCIPLES

In relation to analytical chemistry, photometry refers to the measurement of the light-transmitting power of a solution in order to determine the concentration of light-absorbing substances present within. Photometry can, of course, be applied to measure the transmission of energy in the ultraviolet, infrared, and visible regions of the radiant energy spectrum. Instruments that are used to measure transmittance at various wavelengths are called spectrophotometers or photoelectric colorimeters, depending upon certain essential differences in construction, particularly the method of producing monochromatic light.

In all such instruments, monochromatic light is passed through an absorbing column of an often coloured solution of a fixed depth and directed upon a photosensitive device which converts the radiant energy into electrical energy. The current produced under these conditions is measured by means of a galvanometer or a sensitive voltmeter.

Absorbance as measured in photometers involves not only the absorbance of the solute in the solution being evaluated, but also all of the molecules of the liquid through which the light passes. It is necessary, therefore, to adjust the instrument by means of a "blank." This blank is prepared by placing in the absorption cell (cuvette) all of the constituents of the unknown solution (solvents, reagents, and so forth), but under conditions that will not permit the colour reaction to take place. The blank solution is used to set the meter of the instrument at a fixed point (100% T or zero absorbance). After adjusting the meter, the cuvette containing the unknown is placed in the instrument and read. By this means, the absorbance of the reagents used can be cancelled. It follows that the reading of the meter with the unknown cuvette in place is a measure of the amount of absorbance of

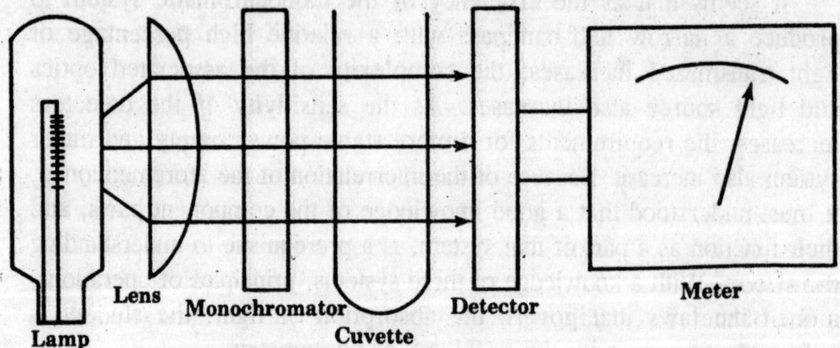

Fig. 13.1. Schematic of single-beam system.

monochromatic light by the unknown. The greater the number of molecules or ions of absorbing substance present, the greater is the absorption of light. In other words, the deeper the colour, the greater is the deflection of the galvanometer from its original setting. Thus, the concentration of absorbing component present in a solution may be accurately measured by a photometer, provided that the necessary monochromatic light is used.

BEER'S LAW

Although the laws of photometry have been referred to as "Beer's Law" (for convenience sake), they are actually laws derived by five different individuals. Bouguer showed (1729) that when light passes through an absorbing body, there is always the same difference between the logarithms of the amount of the light entering and light leaving any section of the same thickness; or, as Lambert expressed it (1760), if one-half the amount of light entering leaves the first section, then ½ × ½ of the original light will leave the second section. Beer (1852) found that one solution of copper sulfate will absorb the same amount as another if the concentration of the first is twice that of the second and the length of the light path of the first is one half that of the second.

These laws then state that light, in passing through a coloured medium, is absorbed in direct proportion to the amount of the coloured substance in the light path. Thus, the strength of the observed "colour" is directly proportional to the concentration of the absorbing chromagen in solution.

If we let P equal the transmitted energy and Po the incident energy, then the ratio P/Po represents the transmittance (T) of the absorbing material. If the material does not absorb at all, then P and Po are the same, and P/Po=T, and %T=100. By using the negative logarithm of T, measurements are transformed to the energy absorbed 1/T. The equation now becomes A = ¾log T; since 2 is the log of 100, the % T formula becomes A=2¾log %T. In practice, we adjust the photometer to read 100% T with a cuvette containing a blank solution. We then substitute a standard or unknown solution and read the meter. Thus, we may use % T and semilog paper or A with regular (Cartesian) graph paper to get a straight line when plotting against concentration as the abscissa. When a straight line is obtained on one of the above graphs, we may also calculate the unknown from the formula: Concentration of the unknown equals A of unknown divided by A of standard times the concentration of the standard.

$$C_\mu = A_\mu / A_s \times C_s$$

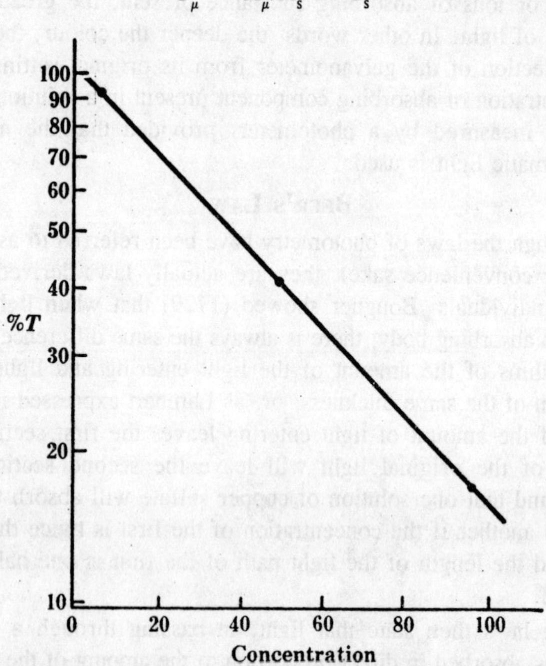

Fig. 13.2. Calibration curve: %T vs concentration plotted on semilog paper.

Component Parts

Power Supplies

The commercial power supply is adequate for most general uses and the usual fluctuations of voltage and current do not significantly upset the equilibrium of lamps, heaters, or motors.

However, no stable operation of light-measuring instruments is possible without better control of the power source. Therefore, a power supply capable of furnishing adequately regulated electrical energy for the particular need of the instrument must be utilized. These power supplies fall into three general categories: batteries, voltage-regulating transformers (Sola), and electronic power supplies.

Batteries, either wet cell or dry cell, produce quite stable voltages and are relatively inexpensive. Dry cell batteries are mainly used for the operation of some small photometers. Wheatstone bridges, and radiation detectors, or wherever portability is necessary and a low direct current supply can be used.

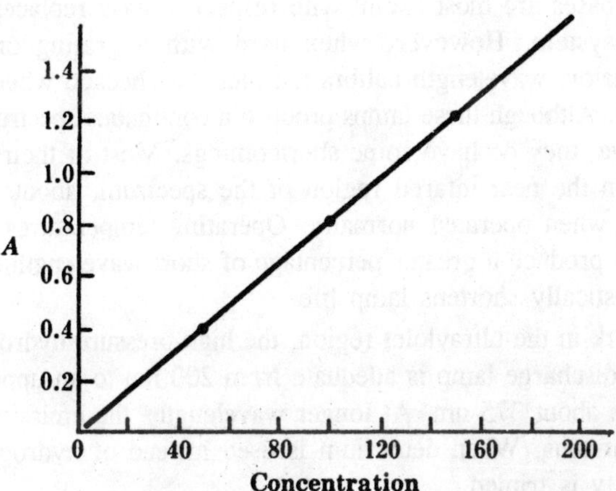

Fig. 13.3. Calibration curve: A vs concentration.

The wet cell (lead-acid) battery, as used in automobiles, also produces a fairly stable current supply. However, it is also limited to being a source of low voltage and direct current. In addition, it requires regular maintenance consisting of frequent recharging and addition of water. After charging, the battery must go through a short period of discharging before stable current is obtained.

The constant-voltage transformers (such as Sola) will regulate power output very closely as long as the input remains between 95 and 125V and at 60 Hz (cps). These devices are essentially trouble-free and have no moving parts.

The electronic power supplies will give more precise regulation than can be easily obtained from the "Sole-type" transformer. Situations may be encountered where frequency variations must be taken into account. This necessitates the use of an electronic device to regulate and supply the desired voltage. These devices are very complex and quite expensive. They utilize either vacuum tube or solid-state regulators. The latter are becoming much more common and it is claimed that they are more dependable. However, it should be remembered that just about anyone can change a vacuum tube, but no inexperienced person should even try to tinker with solid-state components.

Radiant Energy Sources

The function of the light source is to provide incident light of sufficient intensity for measurement. For work in the visible, near infrared, and near ultraviolet regions, the most common source is the glass-enclosed, tungsten filament, incandescent lamp. These lamps with prefocused bases are most useful with respect to easy replacement in an optical system. However, when used with a grating or prism monochromator, wavelength calibration must be checked when lamps are changed. Although these lamps produce a continuous spectrum over a wide range, they do have some shortcomings. Most of their energy is emitted in the near-infared region of the spectrum, about 15% in the visible, when operated normally. Operating temperatures can be increased to produce a greater percentage of short-wavelength energy, but this drastically shortens lamp life.

For work in the ultraviolet region, the high-pressure hydrogen (or deuterium) discharge lamp is adequate from 200 nm to an upper limit extending to about 375 nm. At longer wavelengths the emission is no longer continuous. When deuterium is used instead of hydrogen, the light intensity is tripled.

For very high levels of ultraviolet illumination, the xenon arc or high-pressure mercury vapour lamp provide a large amount of continuous radiation plus high energies at the spectral lines of these elements. These lamps become very hot in operation and may even require thermal insulation, with or without auxiliary cooling, to protect the surrounding components.

Although these special lamps are available and can be used, they usually require special power supplies and mountings. When planning to use these types of energy sources, it is better to purchase the equipment as a unit so that the light source will be properly set up and the necessary insulation will be in place.

Monochromator

When measuring the absorption or emission of radiant energy by an absorbing solution, it is necessary to be able to isolate the desired wavelength of that energy and exclude the rest. In other words, by restricting the band of wavelengths passing through the sample to those absorbed by the substance of interest, the sensitivity of the instrumental measurements to concentration changes is greatly enhanced. Thus the important characteristics of a monochromator (dispersing device) are its bandpass width, the nominal wavelength, and peak transmittance.

There are numerous ways of isolating the desired spectrum. The simplest device is a *filter* (glass, Wratten, or interference) placed just in front of the sample holder. The usual glass and Wratten filters are of relative wide bandwidths and low peak emission. Since the function of the filter is to absorb unwanted energy, the dissipation of the heat produced in the filter must be considered.

Narrower bandwidths are obtained with *interference filters*. This type consists of an evaporated coating of a transparent dielectric spacer of low refractive index sandwiched between semitransparent silver films. Sharp cutoff filters are placed on each side of these films to eliminate other than first-order effects. These filters have a bandwidth of 10-17 nm and a peak transmittance of 40-60%.

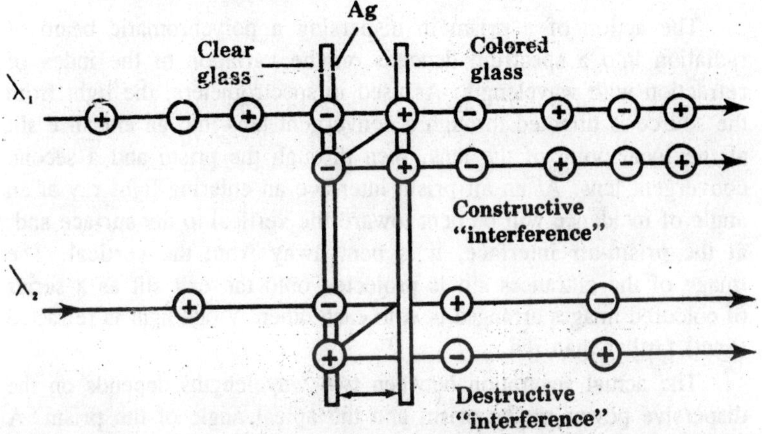

Fig. 13.4. Interference filter.

Multilayer interference filters consist of successive layers of high and low refractive index dielectrics on alternating layers. These filters are characterized by a bandpass width of 8 nm or less and a peak transmittance of 60-90%.

These multilayer interference filters will transmit only light with a wavelength two times the distance between the silver films (in phase), that is, with complete constructive interference. All other wavelengths will show partial or complete destructive interference as the reflected light rays are retransmitted out of phase with those of the same wavelength which are transmitted directly.

These interference filters can be used with high-intensity light sources since they remove unwanted radiation by transmission and reflection and not by absorption.

Many instruments use *prisms* or *diffraction gratings* as monochromators. These devices separate the various wavelengths of radiant energy as produced by a tungsten lamp by refraction or diffraction, and present them as a spectrum from which the desired wavelengths may be selected.

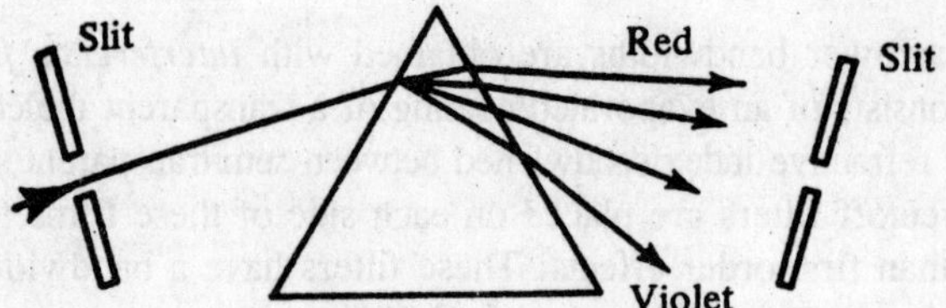

Fig. 13.5. Light dispersion by prism.

The action of a prism in dispersing a polychromatic beam of radiation into a spectrum depends on the variation of the index of refraction with wavelength. As used in spectrometers, the light from the source is directed through a convergent lens into an entrance slit at the focal point of the lens, then through the prism and a second convergent lens. At an air-prism interface an entering light ray at an angle of incidence will be bent toward the vertical to the surface and, at the prism-air interface, it is bent away from the vertical. The image of the entrances slit is projected onto the exit slit as a series of coloured images arranged next to each other. Violet light is refracted (bent) farther than red.

The actual separation between two wavelengths depends on the dispersive power of the prism and the apical angle of the prism. A nonlinear wavelength scale results. The longer wavelengths are not

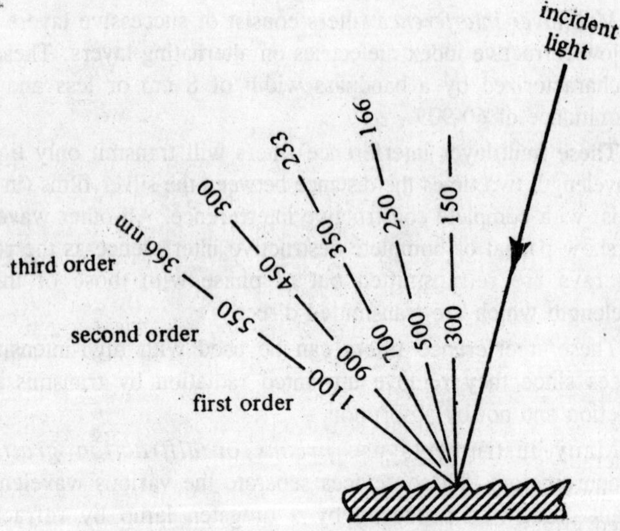

Fig. 13.6. Light dispersion by diffraction grating, showing possible second and third order interference.

refracted as much and thus are crowded. Dispersion by glass is about three times that of quartz. Quartz or fused silica is mandatory for inclusion of the spectra below 350 nm.

Diffraction gratings, on the other hand, produce a linear noncurved spectrum. A diffraction grating is based on the principle that light rays of radiant energy will bend around a sharp corner, the degree of bending depending on the wavelength. Thus, with a beam of energy containing many wavelengths striking a grating, many tiny spectra are produced, one for each line of the grating. As the wavelengths move past the corners, wavefronts are formed. Where these cross, those that are in phase reinforce one another, and those that are not cancel out and disappear, thus leaving a complete spectrum to display upon the exist slit, much the same as with the prisms. Some grating monochromators have a half-bandwidth as high as 35 nm, while other more expensive units have half-band-widths of less than 0.5 nm.

Despite the good dispersive qualities and the linear spectrum presentation, diffraction gratings have problems with stray light. The stray light may be defined as radiant energy at unwanted wavelengths reaching the detector. One problem is slight imperfections in the ruling. Another is caused by second-order effects of the grating. Part of this trouble can be eliminated by using a double monochromator or by a special mounting such as the Ebert system.

Sample cell (cuvette)

Absorption spectrophotometry usually evaluates the absorption of a solute in a liquid solvent. The square or rectangular absorption cells have plane parallel faces and a light path of constant length. They are free of optical aberrations. Sets of cuvettes may be matched to very close tolerances. This type of cuvette is rather expensive and is usually utilized in the more expensive instruments. For ultraviolet work (below 340 nm) it is mandatory that this type of cell, constructed from silica or quartz, be used.

For most work in the visual range, the cylindrical test tube type is sufficiently accurate. These cuvettes are prone to variable reflection and refraction errors as well as lens effect. Glass tubing (from which cuvettes are made) is rarely round and is not polished, so three are considerably more surface irregularities. For this reason, round tube-typed absorption cells require close calibration and segregation into matched groups. Because these cells are generally somewhat oval, it is necessary that they be marked in front for proper light-path orientation.

Another solution to this problem would be to use a flow-through cuvette. Since the same cuvette is in the light path at all times, cuvette errors would be compensated for by the blank. Highly reproducible results can be obtained, provided the cuvette does not become dirty and no bubbles are introduced and/or trapped in the cell. A source of vacuum is needed to empty the cell. Filling is often a very real problem because of the small tops. This may lead to spillage into the instrument and resultant corrosion.

Associated Optics

The range of transmittance of materials for construction of windows and lenses is a critical factor. The absorbance of any material should be less than 0.2 at the wavelength of use. Ordinary silica glass transmits satisfactorily from 350-3000 nm. Special Corex glass will extend the ultraviolet range to about 300 nm. Quartz or fused silica must be used below this.

Beam reduction is accomplished by condensers that can reduce the beam size by a factor of 25 without loss of appreciable energy.

In colorimeters and economically priced spectrophotometers, simple lenses are used to focus or collimate the light beams. In the more expensive instruments, front-surfaced mirrors are used to reduce energy loss. These mirrors are aluminized on their front surfaces are other metallic surfaces show selective absorption at certain wavelengths. The surfaces of these mirrors are coated with a thin film of magnesium fluoride to reduce light scattering.

Detectors

Any photosensitive device may be of use as a detector of radiant energy, provided that it has a linear response in the part of the spectrum to be used and is sensitive enough for the task at hand. These devices must first convert electromagnetic energy to a different type of energy, namely electrical energy. The electrical energy produced can then be measured.

The most common of these is the photocell. These require no external voltage source, but rely on internal electron transfers to produce a current in an external circuit. These cells are composed of an iron back plate and a layer of crystalline selenium or cadmium on one surface. Electrical contacts are made with the iron plate and an electrical conductive transparent film on the front active layer. The cells are then sealed to prevent physical or chemical damage.

A photon striking one of the selenium or cadmium atoms transfers its energy to an electron and raises it into a conduction band. This

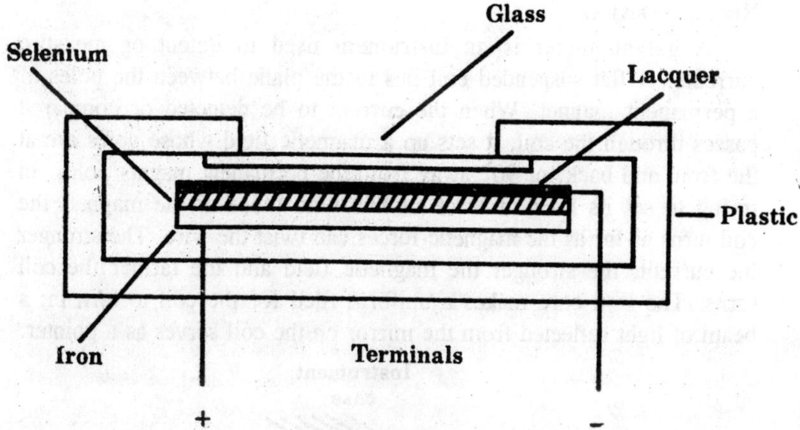

Fig. 13.7. Photocell.

electron now travels from the front to the back of the photocell and through the external circuit for measurement. These cells are simple, quite inexpensive, and in general are very dependable. However, they have certain faults that should be understood. These cells can become "light blinded," a condition similar to that of the human eye. They need time to rest. Their response is temperature sensitive, requiring some warm-up time for temperature stability. Their low internal resistance and low output of electrical energy are not easily amplified.

Phototubes are constructed of a negatively charged cathode and a wirelike positively charged anode. The cathode is coated with a photoemissive substance such as cesium oxide. When a photon strikes their layer, electrons are emitted and jump through the vacuum over to the anode, where they are collected and return via the external circuit.

The output from these tubes can be amplified and then fed into external circuits for greater sensitivity. For this reason, this type of detector is used on all precise instruments that have a close restriction on the slit and thereby the wavelength presentation.

The photomultiplier tube combines photocathode emission with multiple cassade stages of electron amplification of primary photocurrent within the tube itself. The tube is constructed so that the primary photoelectrons from cathode are attracted and accelerated to several succeeding dynodes. These dynodes are constructed of a material that will give off several secondary electrons when hit by other high-energy electrons. Thus the photomultiplier tube can measure light intensities about 200 times weaker than the ordinary phototube.

Readout Devices

A galvanometer is an instrument used to detect or measure currents. A flat suspended coil lies in the plane between the poles of a permanent magnet. When the current to be detected or compared passes through the coil, it sets up a magnetic field whose poles are at the front and back, or 90° away from the permanent magnet poles. In trying to set its lines of force in line with those of the magnet, the coil turns as far as the magnetic forces can twist the wire. The stronger the current, the stronger the magnetic field and the farther the coil turns. The iron core makes a uniform filed for the coil to turn in; a beam of light reflected from the mirror on the coil serves as a pointer.

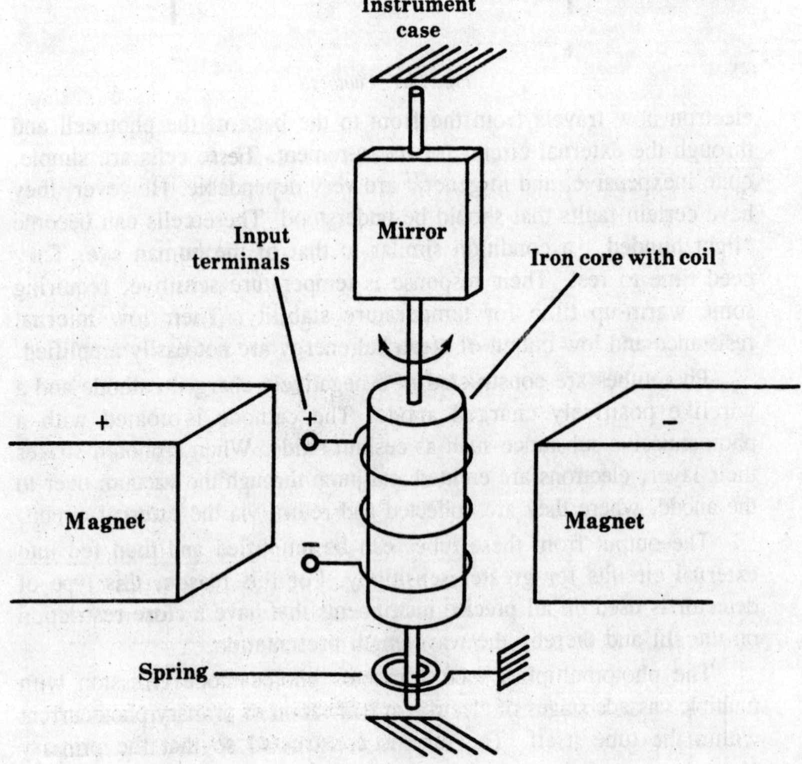

Fig. 13.8. Galvanometer schematic with terminals.

When an arbitrary readout device, such as percent scale or optical density scale, is incorporated into this system, we have what is called a *direct readout meter*. Actually, this readout could be in millivolts, milliamps, mg%, mEq/liter, or any other arbitrary unitage, provided the scales take into consideration the linear relation of milliamps and % T and the log relationship between millivolts and absorbance. Therefore, this type of readout is generally considered to be the fastest, simplest, and the easiest to use.

If, on the other hand, an electrical force governed by the action of a variable resister (potentiometer) is used to bring the meter to null point, the system is said to be a *null-point system*. In this case, an arbitrary scale is fitted to the potentiometer scale. This scale may carry the same arbitrary type of unitage as the first.

This slide wire potentiometer may also be connected to a servomotor of a digital readout system or of a recorder.

Types of Instruments

Although the filter photometer is not to be considered as a spectrophotometer, because it does not give a continuous source of monochromatic radiant energy, they must be considered together when it comes to function and system design. Colorimeters that use the regular glass filter will also use a simple low-energy light source. In fact, this type of filter dictates the use of this type of source, as well

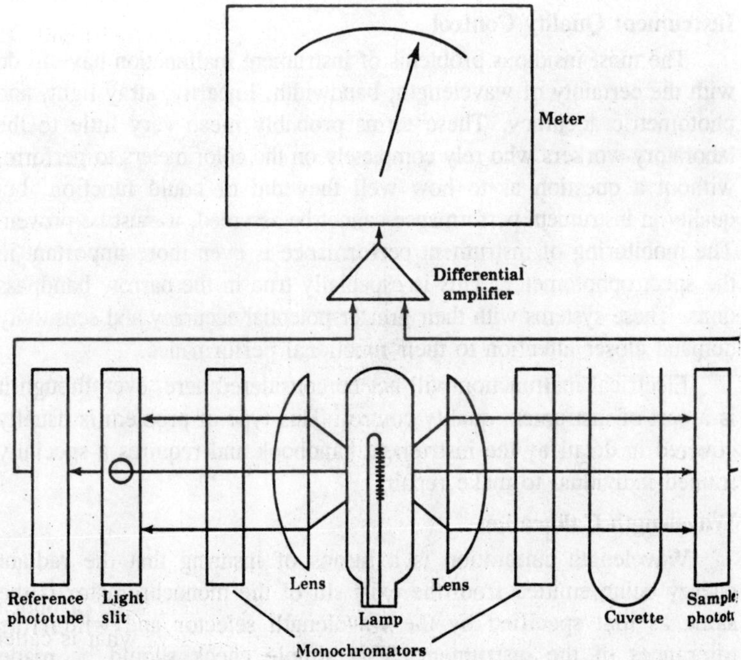

Fig. 13.9. Double-beam system with optics.

as eliminating the need for even a single lens. Therefore, the simplest system would require a power supply, light source, filter, sample holder, detector, and a meter in that order. However, when an interference filter, prism, or diffraction grating is use, a more elaborate system is necessary. These types of monochromators require that radiant energy be focused into the system in parallel lines. Diffraction gratings and prisms also require the use of entrance and exit slits to *limit* the spectrum and exclude stray light. In fact, most grating monochromators require that sharp cutoff filters be placed at the entrance to eliminate second-order spectrum. When you add to this a sample holder, a detector, and a meter, you have then a single-beam spectrophotometer.

In a double-beam system monochromatic light from either a single or double monochromator is focused through both a reference and a sample compartment. The intensity of these two beams of light is then measured by either one or two detectors, and the sample beam is compared to the reference side as a ratio. This ratio may then be fed into a meter or directly into a ratio recording and a beam splitting and reuniting system of choppers to overcome the problem of balancing two phototubes to an exactly equal spectral response.

Instrument Quality Control

The most insidious problems of instrument malfunction have to do with the certainty of wavelength, bandwidth, linearity, stray light, and photometric accuracy. These terms probably mean very little to the laboratory workers who rely completely on the colorimeters to perform, without a question as to how well they did or could function; but quality in instrument performance cannot be assumed, it must be proven. The monitoring of instrument performance is even more important in the spectrophotometer. This is especially true in the narrow bandpass units. These systems with their greater potential accuracy and sensitivity demand closer attention to their functional performance.

Electrical malfunction will not be considered here, even though it is a part of instrument quality control. This type of problem is usually covered in detail by the instrument handbook and requires a specially trained individual to make repairs.

Wavelength Calibration

Wavelength calibration is a means of insuring that the radiant energy being emitted from the exist slit of the monochromator is the same as that specified by the wavelength selector and within the tolerances of the instrument. This simple check should be made whenever a new lamp is installed and routinely thereafter. This is to insure that the slight changes in lamp filament position during usage are corrected and do not lead to gross analytical errors. Minor errors in wavelength settings are of relatively little importance in wide bandpass spectrophotometers, and these units may be checked less often. Wavelength calibration of the more sophisticated narrow bandpass instruments are of the utmost importance, and no scan of any substance is worth doing if the monochromator is not in calibration.

Didymium and holmium oxide glass filters have been used for many years on the general laboratory instruments. More recently, the combination of the didymium glass filter and an acid solution of nickel sulfate has been used. This solution gives a peak transmission at 510 nm. A solution of cobalt chloride has a peak absorbance at this wavelength. This solution may be helpful as a secondary check on the nickel sulfate when it is not in agreement with the didymium filter.

Precision instruments require a more precise means of wavelength calibration. The emission bands of mercury arc or deuterium lamps are good sources of radiant energy with specific emission spectra. The quartz mercury arc lamp, with its multiple emission bands, provides the best source for accurate wavelength calibration from 205—1014 nm. The deuterium lamp in ultraviolet instruments has two major emission lines, 486 and 656.2 nm, that provide a good source of wavelength calibration in the visual range. This lamp provides the most practical means of wavelength calibration of a grating instrument, for only two lines are needed when light dispersion is linear.

Stray Light

Stray light is any radiant energy measured by the detector that is outside the spectral region isolated by the monochromator of the instrument. It can be caused by dirty optics, poor baffling, or faulty grating in the monochromator, as well as fluorescence of the sample itself. Because stray light may be independent of sample concentration, its relative effect is usually greatest at low transmittance, where instrument errors are also greater.

The American Society of Testing Material E-13 Committee Standards have been adopted for the measurement of stray radiant energy in spectrophotometers and provides a good means of instrument evaluation. However, for ease of understanding and routine checking, a common sense approach may be more practical. Scattered light can be of any wavelength, and in grating instruments it may also be of second-order reflection. This may also be true of old and poorly constructed interference filters. Thus, this light may well be of a wavelength far removed from that selected and will not be absorbed by the test solution. From this information, it is easy to reason that scattered light can cause nonlinearity and insensitivity.

Sharp cutoff filters may be used to directly assess the amount of stray light at the extremes of the instrument range. These filters have the unique ability to stop all light above or below a certain range. Thus, by the proper selection of filters, it is possible to check the apparent stray light of any instrument. This may be done by setting the wavelength selector at the desired wavelength (400 or 700 nm) and then adjusting the instrument meter to read 100% transmission. Place a filter in the sample compartment and not the meter reading. This reading is an indication of stray light and must be considered as an instrumental error.

If one is performing time rate analyses, one of the simplest checks to make is to measure the absorbance of NADH at 340 nm with the tungsten and deuterium light source. If the absorbance of the solution is greater with the deuterium lamp, stray light was present when the tungsten lamp was used and this has been corrected by employing the deuterium radiation source. This is because the deuterium lamp is a discontinuous radiant energy source in the visual range, and therefore does not emit radiation at 680 nm, the wavelength of light that is of primary importance in second-order effect. Also very little energy is emitted by this lamp above 400 nm, so that scattered-light problems are greatly diminished when working in the ultraviolet range.

Nickel sulfate solution has also been used for this purpose over the visual range. This solution shows maximum absorption at 400 and 700 nm, and under certain circumstances no light should be transmitted at these wavelengths. Any deviation from this on repeated checks is an indication of stray light.

Photometric Accuracy

Provided that the stray-light checks is negligible, the linearity check tests the ability of the photocell to produce a signal proportional to the light intensity and of the instrument's meter system to measure this signal accurately. A linearity check can be made by reading the absorbance of standard solutions of various salt solutions that have major absorption peaks at certain wavelengths or by using neutral density filters that absorb energy over a broad range.

Solutions of potassium chromate and potassium dichromate have been used for evaluating photometric accuracy in the ultraviolet range. A solution of 0.04 g/liter of potassium chromate dissolved in 0.05N potassium hydroxide will show maximum absorbance at 273 and 373 nm. The absorbance range reported for this solution is 0.189—0.199A and 0.247—0.249A, respectively, for these two wavelengths. When measuring absorbance below 260 nm, this solution should not be more than 6 months old. This may be the reason for the wide range in absorbance values at 273 nm. A solution of potassium dichromate,0.05 g/liter dissolved in 0.01N sulfuric acid, also exhibits two absorbance peaks that may be used. These are at 257 and 350 nm and should read 27.6% transmission at the latter wavelength. Absorbance patterns of this solution are varied by changes in acidity: therefore, the pH should be carefully controlled. Cobalt ammonium sulfate and Thompson's solution have been used for this evaluation in the visual range. These solutions require great care in preparation and handling. The absorbance cells used must be cleaned and matched before use. Although these materials are relatively stable, they may react with contaminants and deteriorate. The solvent solution must also be checked for absorbance in the ultraviolet range when these solutions are to be analyzed. Frings, et. al., has proposed using French's green food colouring for monitoring detector response at three wavelengths. This dye solution has absorbance maxima at 257 nm, 410 nm, and 630 nm. The advantage of this solution is that it provides a means of response verification in both visual and ultraviolet ranges with one solution at a very modest cost.

The National Bureau of Standards developed a set of neutral density glass filters for the specific purpose of checking the photometric scale of spectrophotometers. These filters are premounted in holders ready for use in the standard 10-mm cell compartment. They are not readily affected by temperature change nor do they require a critical wavelength adjustment. These standard reference materials (SRM) come as a set of three individual filters calibrated and certified to ± 0.5% transmittance over four different wavelengths of the visual range. These filters were the forerunners of other sets manufactured by Bausch and Lomb as well as Chemetrics Corporation. All of the later filter sets have a filter that will fit a standard 10-mm cell compartment. These sets provide a means of checking for stray light, bandpass, and wavelength as well as photometric accuracy.

14

CYTOLOGICAL TECHNIQUES

The study of cells has assumed much significance for biology students as well as research scholars and it is no longer enough just reading about it. Students specializing in cell biology are actually expected to follow various sophisticated techniques. Some of the simple techniques described in the following pages are concerned with the cell organelles and these are essential in staining them for both qualitative and quantitative recordings.

STUDY OF MITOCHONDRIA

Any tissue can be selected for the study of mitochondria. The tissue is generally fixed in neutral formalin for 24 hours at room temperature followed by post-chroming for 3-24 hours at 37°C or in Legaud's fixative (4 vols. of 3% aqueous potassium dichromate and 1 vol. of 4% formaldehyde, mixed just before use) for 4-24 hours at 8°C followed by post-chroming as in neutral formal and washing overnight in running tap water.

Fuchsin Acid-Light Green Method: (A Rapid Method for Staining Mitochondria)

Solutions required

A. Hydrochloric acid N/1

B. Fuchsin acid, 1% aqueous

C. Light green SF, 1% aqueous

Procedure

1. Paraffin sections of formalin-fixed material are mounted on slides and dewaxed with xylol as usual.
2. Pass the slides through the usual graded alcohols into water.

3. Hydrolyse for three minutes in the hydrochloric acid (solution A), at 60°C.
4. Rinse in water.
5. Stain in the fuchsin acid solution for thirty seconds.
6. Rinse in water.
7. Counterstain in the light green solution for one to three minutes.
8. Rinse in water.
9. Dehydrate with 95% and absolute alcohol.
10. Clear in xylol.
11. Mount in D.P.X. or in Canada balsam in xylol.

Results

The mitochondria stand out sharply stained purplish red with green peripheral wall. Chromatin and collagen are green. Muscle tissue: purplish. Erythrocytes: brilliant red.

Note

It may be interest to compare the results obtained with this technique with those obtained using the same dyes, but under simpler conditions and without hydrolysis, in the techniques of *MacConaill* and *Gurr*.

Fuchsin Acid (Altmann)—Methyl Green Method (for Mitochondria in Leucocytes)

Solution required

A.	Potassium dichromate	15 gm
	Distilled water	100 ml
B.	Iodine 0.5% in 70% alcohol	
C.	Sodium thiosulphate 0.5% aqueous	
D.	Altmann's acid fuchsin	
E.	Methyl green, 1% aqueous	

Procedure

1. Make smears of fresh, unoxalated blood on scrupulously clean slides, and fix by placing them at once in Helly's fluid and leaving therein for twenty-four hours.
2. Post-chrome in solution A for forty-eight hours.
3. Immerse in solution B for two minutes.
4. Rinse in 70% alcohol.
5. Immerse in the sodium thiosulphate solution for one minute.

6. Wash in running water for several hours or overnight.
7. Transfer to distilled water.
8. Take a slide from the distilled water and blot or wipe away the water from all parts of the slide except that covered by the smear.
9. Place the slide over the corner of a tripod, flood with solution D and heat the preparation over a flame until fumes appear.
10. Pour off excess stain and wash with water.
11. Stain for a few seconds with methyl green by allowing a few drops of the solution to flow over the slide.
12. Dehydrate in two changes of absolute alcohol.
13. Clear in xylol and mount in Cristalite or D.P.X.

Results

Mitochondria, erythrocytes, and the specific granules of eosinophilic leucocytes are stained red. Nuclei are stained by the methyl green, while the cytoplasm also takes this dye but to a lesser extent.

Notes

(a) The use of oxalated blood is not to be recommended due to the alteration produced, by the commonly used oxalates, in the cytoplasm and nuclear structure of the leucocytes.

(b) Fixation of the blood smears before they are dry is absolutely essential, according to the author, to obtained well differentiated preparations.

Fuchsin Acid—Pieric Acid Method (After Altmann)

Solutions required

A. *Altmann's Fluid*

Potassium dichromate 5% aqueous	1 volume
Osmic acid 2% aqueous	1 volume

Note: Although the penetration power of this fixative is poor, it is very satisfactory for surface fixation.

B. Acid fuchsin	20 gm
Aniline water	95 ml

C. Picric acid saturated in absolute alcohol.

Procedure

1. Small pieces of tissue, not more than 2 mm. in diameter, are fixed for twenty-four hours in Altmann's Fluid.
2. Wash for an hour in running water; then dehydrate; clear, and embed in paraffin wax in the usual manner.

3. Sections, not thicker than 4 μ, are brought down to distilled water; then stained for six minutes in Solution B.
4. Pour off excess stain; then blot section carefully; then differentiate and counterstain by flooding the preparation with Solution C.
5. Rinse quickly in 95% alcohol; then dehydrate with absolute alcohol; clear in xylol, and mount.

Results

Mitochondria are stained crimson against a vivid yellow background.

Fuchsin Acid—Toluidine Blue—Aurantia Method

Solutions required

A. *Champy's Fluid*

Potassium dichromate 3%	7 ml
Chromic acid 1%	7 ml
Osmic acid 2%	4 ml

B. Pyroligneous acid — 1 volume
Chromic acid 1% — 2 volumes

C. Potassium dichromate 3%

D. Acid fuchsin — 10 gm
Aniline water — 100 ml

E. Toluidine blue 0.5% aqueous.

F. Aurantia 0.5% in 70% alcohol.

Procedure

1. Tissues are fixed in Champy's fluid for 24 hours; then washed in running water for at least one hour.
2. Immerse in solution B for 12-24 hours, then wash in distilled water for 30 minutes.
3. Mordant in solution C for three days: then wash in running water for 24 hours.
4. Dehydrate; clear; embed in paraffin wax.
5. Sections, 3-5 μ in thickness are taken down to water in usual manner; then stained for 6 minutes by flooding the slide with solution D and heating gently till vapour rises.
6. Rinse in distilled water; then counterstain solution E.
7. Rinse with distilled water; then stain with solution F for 30-50 seconds.
8. Differentiate with 95% alcohol; dehydrate; clear in xylol and mount.

Results

Mitochondria are stained red; nuclei-blue; background yellow.

Phospho-Tungustic Acid Hematoxolin (PTAH) Method (Mallary)

Solutions required

1. Lugol's iodine solution
2. Aqueous solutin of sodium thiosulphate 0.5%.
3. $KMnO_4$ aq. solution—0.25%
4. Aq. Oxalic acid solution—5%
5. Phospho-tungustic acid Hematoxylin

Hematoxylin	1 gm
Phospho-tungustic acid	20 gm
Distilled water	1000 ml

Dissolve the solids in separate portions of water, the hematoxylin with the acid of gentle heat. When cooled, combine. No preservative is necessary, spontaneous ripening requires several weeks. It can be accomplished at once by adding 0.177 gm of $KMnO_4$.

Procedure

1. Paraffin sections of Helly fixed material are brought to water through toluene and alcohols and placed in a half strength Lugol's iodine solution until yellowish brown. Wash off excess iodine in water and dip in and out of 0.5% aqueous solution of sodium thiosulphate until the section is colourless. Wash several minutes in running tap water.
2. Place sections in a freshly made up 0.25% aqueous solution of $KMnO_4$ for 5 minutes.
3. Wash in water.
4. Put in oxalic acid 5% aqueous solution for 1 ml or until clear and colourless.
5. Wash in running tap water for several minutes.
6. Stain in phosphotungustic acid hematoxylin for 12-24 hrs. Overnight staining (15 hrs) is found quite adequate.
7. Blot sections dry with paper towel or filter paper and pass rapidly through 95% and absolute alcohol, agitating the slides and shortening the time as much as possible since alcohol will extract the red part of the stain.
8. Clear in xylol and mount.

Results

Mitochondria of liver and kidney cells stained deep blue.

PAS-PTAH Method

Solution required

1. Regaud's Fixative;

   3% aqueous Pot. dichromate 80 ml

   Formalin 20 ml
2. 3% Pot. dichromate solution.
3. Altmann's Aniline Acid Fuchsin

   Aniline 1 ml

   Acid Fuchsin 4 gm

   Distilled water 20 ml

   Shake aniline with distilled water for a couple of minutes Le stand and then shake again several times at intervals over 24 hrs. Filter. Add 4 gm of acid fuchsin and shake at intervals over several hours. The solution keeps only long enough to be used the same day it is made.
4. Dilute sodium carbonate 0.1% aqueous.
5. 1% Hydrochloric acid.
6. 0.5% aqueous methyl blue.

Procedure

1. Fix tissue in Regand's fixative for four days, making up new Regaud's each day. Post chrome for 8 days in 3% Pot. dichromate changing every day.
2. Wash in running water for 24 hrs.
3. Dehydrate in graded alcohol, clear in toluene and mount in paraffin. Cut paraffin sections, 2.4 μ thick.
4. Bring paraffin sections to water and place in steaming Altmann's anilins acid fuchsin for 5-10 minutes removing the heat from the solution, when the slides are placed in it.
5. Differentiate in dilute sodium carbonate until cytoplasm in pale pink and nearly colourless.
6. Stop differentiation and heighten colour by dipping briefly in 1% HCI.
7. Wash in distilled water and counterstain in 0.5% aqueous methylene blue. This will take only a few seconds.
8. Rinse in distilled water, dip in 1% HCI briefly.
9. Wash in distilled water, dehydrate in alcohol.
10. Clear in toluene and mount in balsam.

Results

The mitochondria become a bright red nuclei become blue to green coloured.

Van Gieson's Stain Method

Solution required

1. Van Gieson stain:

Acid Fuchsin 1% aq.	10 c.c.
Picric acid saturated aq. solution (about 1.22%)	100 c.c.

2. Alcoholic Hematoxylin

Hematoxylin	0.5 gm
95% alcohol	10 c.c.
Distilled water	90 c.c.

Allow this stain to ripen for 4-5 weeks then dilute this with equal parts at distilled water.

Procedure

1. Bring sections to 95% alcohol.
2. Place in alcoholic iodine for 5 mts. to remove mercuric chloride.
3. Rinse in 80% alcohol or wash in tap water.
4. Transfer to 5% sodium thiosulphate for 2 minutes to bleach iodine.
5. Wash well in running tap water.
6. Mordant in ammonia ferric alum (2.5%) solution for 1—2 hrs.
7. Wash quickly in distilled water.
8. Stain for 1—36 hrs in alcoholic hematoxylin.

   Note. The section should be adequately stained when on being taken out of the hematoxylin, they are a homogenous jet black and microscopically show no cellular detail at all.
9. Wash in water.
10. Differentiate in 2.5% Iron alum. The sections should be rinsed in tap water before each examination, which will immediately stop the decolourization.
11. Wash in running water 15—60 mts.
12. If desired counterstain with Van Gieson's stain or eosin.
13. Rinse with 50% alcohol.
14. Dehydrate in two changes of 95% alcohol.
15. Two changes in 100%, three in xylol and then mount.

Results

Mitochondria, chromatin, centrioles, nuclei are stained black.

Study of Golgi Complex

The Golgi complex is studied in sections by using different techniques. The tissues which can be easily handled for Golgi preparation are epididymus, small intestine, brain, spinal cord, pancreas. Some of the staining methods for Golgi complex are given below:

Silver-Nitrate—Gold Chloride—Paracarmine Method

Solution required

A. Cobalt nitrate 1% aqueous — 100 ml.
Formalin — 15 ml.

B. Silver nitrate 1.5% aqueous.
(N.B.: This should be stored in an amber or blue glass bottle).

C. Cajal's Reducer.
Hydroquinone 2% aqueous — 100 ml.
Neutral formalin — 15 ml.
Sodium sulphite anhydrous — 0.5 gm
N.B.: This solution should be freshly prepared.

D. Gold chloride 0.2% aqueous.

E. Sodium thiosulphate 5% aqueous.

F. Paracarmine (Mayer).

Procedure

1. Pieces of tissue no thicker than 3 mm. are fixed from two to eighteen hours in the cobalt nitrate formalin solution, according to the size and nature of the material.
2. Wash the tissue quickly in a large volume of distilled water.
3. Immerse in the silver nitrate solution in the dark for thirty-six to forty-eight hours.
4. Wash quickly in a large volume of distilled water.
5. Trim the tissue to a thickness not exceeding 2 mm.
6. Immerse in Solution C (Cajal's Reducer) for two to twenty-four hours in the dark.

   Note: For most soft tissues about four hours will suffice.
7. Wash in several changes distilled water.
8. Dehydrated, clear and embed in paraffin wax in the usual way.
9. Cut sections up to 8 μ in thickness.

10. Fix sections to slides; dewax and pass through descending grades of alcohol to distilled water.
11. Tone sections on slides by immersing in the gold chloride solution for five to ten minutes.
12. Wash quickly in distilled water.
13. Fix in 5% sodium thiosulphate (Solution E) for ten to fifteen minutes.
14. Wash thoroughly in distilled water.
15. Counterstain with paracarmine for about ten minutes.
16. Rinse with 90% alcohol, followed by absolute alcohol.
17. Clear in xylol and mount.

Results

Golgi apparatus stained black while cells are pink or red.

Gold Chloride Method

Solution required

0.1% aq. yellow gold chloride solution
5% sodium thiosulphate solution
Hematoxylin eosin stain solution.

Procedure

1. Bring the section to water and then to distilled water and placed for 5—10 minutes in 0.1% aq. yellow gold chloride solution containing a few drops of glacial acetic acid per 100 ml.
2. Rinse quickly in distilled water.
3. Fix in 5% sodium thiosulphate for 5—10 minutes.
4. Wash in distilled and tap water, stain with hematoxylin and eosine or other desired method.
5. Dehydrate and mount.

Results

Golgi complex—black, Nuclei—blue.

Osmic Acid Method

Solutions required

1. Flemmings Fluid;

1% chromic acid	–75 ml
1% Aq. osmium tetraoxide	–20 ml
Sodium chloride	0.75 gm

2. Mann's Fixative

   1% Aq. osmium tetraoxide 50 ml

   0.75% Aq. sodium chloride -50 ml

   saturated with mercuric chloride.

3. Osmic acid solution

   1% Aq. osmic acid.

   Note. All these osmic acid solutions produce very irritating vapours and must be kept tightly sealed and away from the eyes and hands.

Procedure

1. Fix very small pieces of tissue in Flemming's or Mann's fluid for 24 hrs.
2. Wash in running water for 6—12 hrs and transfer to several changes of distilled water.
3. Place in 1% aq. osmic acid at 37°C for 4 days.
4. Pour off osmic acid and wash in running tap water for 6—18 hrs.
5. Dehydrate in alcohol or acetone, clear in xylol or tolune and embed in paraffin.
6. Cut sections 4 to 6 microns.
7. Deparaffinize and mount.

Results

Golgi elements—black.

Cajal Method (Daven port)

Solution required

1. In alcoholic Nitric acid solution.
2. Alcoholic silver nitrate solution.

   Silver nitrate 10 gm

   Distilled water 10 ml

   When solution is complete then add

   95% alcohol 90 ml

3. 5% Pyrogallol solution in alcohol

   Pyrogallol -5 gm

   95% alcohol -100 ml

Procedure

1. Fix nervous tissue 2 days or longer in 10—20% formalin.
2. Embed in nitrocellulose, cut sections and place them in 70% or 80% alcohol.

3. Transfer to an approximately in alcoholic solution of nitric acid made by adding slowly and with stirring 15 ml of conc. HNO_3 to 85 ml of 95% alcohol. Incubate for about 2 hrs. at 37—44°C.
4. Pass the sections individually through 3 changes of 80% alcohol, 1—2 minutes in each change, wash out the acid.
5. Impregnate at 37—44°C in darkness for 4—24 hrs in the solution of alcoholic silver nitrate.
6. Rinse in 95% alcohol for 3—5 seconds and reduce in 5% Pyrogalled solution prepared in 95% alcohol. (If this solution is to be kept for several days, add 5 ml of formalin to it to retard its oxidation by air. It reduces equally well with or without formalin).
7. Wash well in running water, dehydrate, clear and mount.

Results

Golgi complex—black.

Study of Nucleic Acids

The nucleic acids—Deoxyribonucleic acid (DNA) and ribosenucleic acid (RNA) are found in all animal and plant tissues, and are usually combined with basic proteins so as to form nucleoproteins. DNA is mainly found in the nucleus, and RNA in the cytoplasm. Hydrolysis of nucleic acids yields the following components: (a) phosphate groups, (b) five-carbon sugars and (c) nitrogenous bases purines and pyrimidines.

The two enzyme techniques for the digestion of nucleic acids are invaluable as controls for the methods discussed. Deoxyribonuclease is specific for the removal of DNA in sections, while not affecting of RNA contents.

When ribonuclease is applied to tissue sections, all the RNA is removed whilst the DNA is unaltered. The digestion methods were introduced by Brachet (1940) as controls for the Methyl Pysonin method. Great care must be taken that the enzymes are of a high purity, because impure enzymes will remove all nucleic acids. Both ribosenuclease and deoxyribosenuclease are expensive reagents. Some of the method are discussed as below:

For DNA

Feulgen Nuclear Reaction Method: (By Feulgen and Rossenbeek)

Reagents required

1. Hydrochloric acid
2. Schiff's reagent

3. Potassium metabisulphite
4. Light green, 1 per cent aqueous

Preparation of solutions

1. n-HCl

Hydrochloric acid, (conc.)	8.5 ml
Distilled water	91.5 ml

2. *Schiff's reagent*
3. *Bisulphite solution*

10 per cent Potassium metabisulphite	5 ml
n-Hydrochloric acid	5 ml
Distilled water	90 ml

Sections

All types

Suitable control sections

Pancreas

Procedure

1. Bring all sections to water	
2. Rinse sections in n-HCI at room temperature	1 min
3. Place sections in n-HCI at 6°C,	
4. Rinse sections in n-HCI at room temperature	1 min
5. Transfer sections to Schiff's reagent	45 min
6. Rinse sections in bisulphite, solution (3)	2 min
7. Repeat wash in bisulphite, solution (3)	2 min
8. Repeat wash in bisulphite, solution (3)	2 min
9. Rinse well in distilled water	
10. Counterstain if required in 1 per cent Light Green	2 min
11. Wash in water	
12. Dehydrate through graded alcohols to xylene and mount	

Results

DNA: red-purple

Cytoplasm: green

Remarks

1. The hydrolysis time is important, and the correct time for the fixative must be used.
2. The n-HCI should be preheated to 60°C.

Naphthoic Acid Hydrazine-Feulgen Method

Reagents required

1. 2-Hydroxy-3-naphthoic acid hydrazine.
2. Absolute alcohol
3. Acetic acid
4. Fast blue B
5. Veronal acetate buffer, pH 7.4
6. *n*-Hydrochloric acid

Preparation of solutions

1. n-Hydrochloric acid

Hydrochloric acid, (conc.)	8.5 ml
Distilled water	91.5 ml

2. NAH solution

2-Hydroxy-3-naphthoic acid hydrazide	50 mg
Absolute alcohol	47 ml
Acetic acid (conc.)	3 ml

3. *Fast Blue B solution*

Fast Blue B	50 mg
Veronal acetate buffer, pH 7.4	50 ml

This solution must be freshly prepared.

Sections

All types

Suitable control sections

Pancreas

Method

1. Bring all sections to water	
2. Rinse briefly in n-HCI	
3. Place sections in n-HCI at 62°C,	
4. Rinse sections in n-HCI at room temperature	1 min
5. Rinse sections in distilled water	1 min
6. Rinse sections in 50 per cent alcohol	1 min
7. Place sections in NAH solution at room temperature	3-6 h
8. Rinse sections in 50 per cent alcohol	10 min
9. Rinse sections in 50 per cent alcohol	10 min
10. Rinse sections in 50 per cent alcohol	10 min
11. Rinse sections in distilled water	1 min

12. Place sections in fresh Fast Blue B solution 3 min
13. Dehydrate through graded alcohols to xylene and mount in DPX

Result

DNA: blue to bluish purple
Protein material: possibly purplish red

For DNA and RNA

Methyl Green-Pyronin Method (Modified by Kurwick)

Reagents required

1. Methyl green
2. Chloroform
3. Pyronin Y
4. 0.1 M Acetate buffer, pH 4.8

Preparation of solutions

1. *Methyl Green*

Methyl green	2 g
Distilled water	100 ml

Dissolve the Methyl Green in the distilled water by stirring well. Pour the solution into a separating funnel. Add 100 ml chloroform and shake well; pour off contaminated chloroform and repeat until no more violet is extracted (about 6-8 washes).

2. *Pyronin Y*

Pyronin Y	2 g
Distilled water	100 ml

3. *Staining solution*

Methyl Green	7.5 ml
Pyronin Y	12.5 ml
Acetate buffer, pH 4.8	30.0 ml

Sections

All types, freeze dried recommended

Suitable control section

Pancreas

Procedure

1. Bring all sections to water
2. Stain in Methyl Green-Pyronin solution, 4-10 min
3. Blot dry

4. Rinse rapidly in absolute acetone
5. Rinse rapidly in 10 per cent acetone in xylene
6. Rinse rapidly in 50 per cent acetone in xylene
7. Rinse in xylene
8. Place sections in fresh xylene and mount in DPX

Results

DNA : green

RNA : red

Remarks

1. This method can be unreliable, the main causes being that some samples of Pyronin Y will not work satisfactorily, or that the Methyl Green is not pure enough.
2. The dehydration through acetone and xylene should be rapid.
3. Triethyl phosphate may be used for dehydration instead of acetone.
4. Methyl Green alone may be used in the manner described above to demonstrate DNA.

Methyl Green-Pyronin Method (Modified by Trevan and Sharrock)

Reagents required

1. 2 per cent Methyl Green.
2. 5 per cent Pyronin Y
3. Acetate buffer pH 4.8

Solution A

2 per cent Methyl Green (chloroform-washed)	10 ml
5 per cent Pyronin Y	17.5 ml
Distilled water	250 ml

Solution B

Acetate buffer pH 4.8

Working solution

Solution A	25 ml
Solution B	25 ml

Procedure

1. Bring sections to water
2. Rinse in distilled water and blot dry
3. Stain in working solution 20-30 min
4. Rinse rapidly in distilled water and blot dry

5. Dehydrate in acetone
6. Rinse in acetone xylone 50 : 50, clear in xylene
7. Mount in DPX

Results

DNA : green, bluish green

RNA : red

Remarks

1. Fixation should be in neutral fixatives

Gallocyanin—Chrome Alum Method (By Einarson)

Reagents required

1. Chrome alum
2. Gallocyanin
3. Distilled water

Preparation of solution

Chrome alum	5 g
Distilled water	100 ml
Gallocyanin	150 mg

The chrome alum is dissolve in the distilled water, the gallocyanin added and the solution slowly heated until it boils. It is allowed to boil for 5 min. When the solution has cooled to room temperature, the volume is adjusted to 100 ml. The solution is filtered before use.

Sections

All types

Suitable control sections

Pancreas

Procedure

1. Bring sections down to water
2. Stain in gallocyanin-chrome alum solution 18-48 h
3. Wash in tap water
4. Dehydrate through graded alcohols and mount in DPX

Results

RNA, DNA : blue

Remarks

This method does not distinguish between RNA and DNA but is specific for nucleic acids.

Acridine Orange Method

Solution required

1. Acridine Orange
2. 0.2 M Phosphate buffer
3. Calcium chloride
4. Distilled water
5. Acetic acid

Preparation of solutions

Acridine Orange solution

Acridine Orange	50 mg
Distilled water	40 ml

The pH of the solution is adjusted to 6.0 with phosphate buffer. The volume is then made up to 50 ml.

1. Phosphate buffer, pH 6.0
   See Buffer Tables.
2. Calcium chloride

Calcium chloride	11 g
Distilled water	50 ml

Sections

Frozen sections
Cryostat sections
Paraffin sections — Not formalin-fixed
Freeze dried sections
Smears

Suitable control sections

Pancreas

Procedure

1. Bring all sections to distilled water	
2. Rinse briefly in 1 per cent acetic acid	15 s
3. Rinse sections in distilled water	15 s
4. Stains in Acridine Orange solution	10 sec.—2 min
5. Transfer sections to phosphate buffer, pH 6.0	1 min
6. Differentiate sections in calcium chloride solution	20 s
7. Transfer sections to phosphate buffer, pH 6.0	10 s
8. Mount sections wet and examine under fluorescent microscope	

Results

RNA : red

DNA : light green

Remarks

The authors state that formalin fixation should not be employed

DNA: Extraction (Brachet)

Reagents required

1. Deoxyribonuclease
2. 0.2 M Tris buffer, pH 7.6

Preparation of solution

Extraction of solution

0.2 M Tris buffer, pH 7.6	10 ml
Distilled water	50 ml
Deoxyribonuclease	10 mg

Method

1. Bring test and control sections to water
2. Place test section in extraction solution, control in tris buffer, pH 7.6 at 37°C, for 4 hrs.
3. Wash in running tap water
4. Stain by method (Feulgen method) both sections

Results

Test section : DNA negative

Control section : DNA red

RNA : Extraction (Brachet)

Reagents required

1. Ribonuclease
2. Distilled water

Preparation of solution

Ribonuclease	8 mg
Distilled water	10 ml

Procedure

1. Bring test and control slides to water
2. Place test slide in ribonuclease solution, control slide in distilled water at 37°C for 1 hr

3. Wash in distilled water
4. Apply Method—Methyl Green-Pyronin

Results

Test slide : RNA negative, DNA green

Controlled slide : RNA red, DNA green

Remarks

Nucleic acids may also be extraced by using perchloric acid. This is not as specific as enzyme digestion, but is a reliable technique and may be used on many occassions where extraction is required as considerable reduction is compared with enzyme digestion, the technique is given below.

Extraction of Nucleic Acid with Perchloric Acid

Reagents required

Perchloric acid

Sodium carbonate

Preparation of solutions

1. Perchloric acid	2.5 ml
Distilled water	47.5 ml
2. Perchloric acid	5 ml
Distilled water	45 ml
3. Sodium carbonate	1 g
Distilled water	100 ml

Procedure

To remove RNA only

1. Bring sections down to water
2. Place sections in 10 per cent perchloric acid (solution 2) at 4°C overnight
3. Briefly rinse in distilled water
4. Transfer to the sodium carbonate 5 min
5. Wash in tap water
6. Employ nucleic acid method

Remarks

To remove both RNA and DNA place sections in 5 per cent perchloric acid (solution 1) at 60°C for 30 min at stage (2) then continue method.

Staining Methods for Mast Cell Granules

Dahlia Acetic Method

Solution required

Distilled water	100 ml.
Absolute alcohol	50 ml.
Glacial acetic acid	12.5 ml.
Dahlia	10 gm.

Dissolve by heating in a flask, lightly plugged with cottonwool, on a water bath. Allow to cool; then filter.

Procedure

1. Fix tissues in absolute alcohol and embed in Celloidin.
2. Immerse sections in the staining solution for twelve hours.
3. Differentiate in 95% alcohol.
4. Clear in origanum oil and mount in balsam or in cristalite.

Results

Granules of mast cells are stained reddish violet.

Chryoidin Methods

(For selective staining of mast cell granules)

Method 1—Solution required

Chrysoidin Y, 0.5% aqueous

Procedure

1. Remove paraffin wax from sections with xylol, and take down to 80% alcohol in the usual way.
2. Stain in the chrysoidin solution for five to ten minutes.
3. Rinse in distilled water.
4. Dehydrate in 96% alcohol and two changes of absolute alcohol.
5. Clear in xylol, and mount in Canada balsam in xylol or in a synthetic mountant such as D.P.X. or Cristalite.

Results

Mast cell granules: deep brown to black in type I cell; brown to yellow in type II cell. Nuclei and other tissue elements; slightly yellowish. Glands in subcutaneous tissues: yellow.

Note. The author states that in using simple aqueous solutions of various basic dyes in investigating the staining of tissue containing acid polysaccharides, it was found that only chrysoidin stained the mast cell granules selectively.

Method 2—Chrysoidin-eosin-light green-haematoxylin

Solution required

A. Alum haematoxylin (Carazzi)

Haematoxylin	0.1 gm.
Potash alum	5 gm.
Distilled water	80 ml.
Glycerine	20 ml.
Potassium iodate	0.02 gm.

B. Light green SF, 1% aqueous 1 ml.

Distilled water 49 ml.

Or

C. Eosin, 1% aqueous 0.5 ml.

Tap water 49.5 ml.

Note. Solution C contains 1 : 10,000 of eosin and this may be further diluted, to as 1 : 50,000, with tap water, if desired.

D. Chrysoidin Y, 0.5% aqueous

Procedure

1. Take sections down to 80% alcohol as usual.
2. Stain in the alum haematoxylin for five to ten minutes.
3. Wash in tap water.
4. Stain for three minutes in either solution B or C.
5. Rinse in water.
6. Stain in the chrysoidin for five to ten minutes.
7. Rinse in water.
8. Dehydrate with 96% alcohol and two changes of absolute alcohol; then clear and mount.

Results

Mast cell granules: brown. Nuclei: greyish. Epidermis: slightly yellow, Glands: yellow. Connective tissue: green or red.

Method 3—Chrysoidin-periodic acid-Schiff

Sodium required

A. Periodic acid, 0.7% acqueous

B. Schiff's reagent

C. Sodium bisulphite, 0.52% aqueous.

D. (Optional) Alum haematoxylin (Carazzi)

E. Chrysoidin Y, 0.5% aqueous

Procedure

1. Immerse in solution A for ten minutes.
2. Rinse in water.
3. Immerse in the Schiff's reagent for fifteen to thirty minutes.
4. Immerse for one minute in each of three changes of the sodium bisulphite solution.
5. Wash in running water for ten minutes.
6. (Optional) Stain in the haematoxylin solution for five to ten minutes.
7. Wash in water.
8. Stain in the chrysoidin solution for ten minutes.
9. Rinse in water.
10. Dehydrate with 96% alcohol and two changes of absolute alcohol.
11. Clear in xylol and mount in Canada balsam in xylol or in a neutral synthetic medium.

Results

Mast cell granules: brown. Nuclei: grey. Epidermis: slightly yellow. Glands: yellowish. Connective tissue: intense purple.

In type II cells of abdominal fluids, the granule or its halo of the outer layer is stained weakly with the periodic acid-Schiff procedures.

Staining Method for Negri Bodies

Carbol Aniline Fuchsin Method

Solution required

A. Basic fuchsin	0.5 gm.
Distilled water	80 ml.
Absolute alcohol	20 ml.
Aniline oil	1 ml.
Phenol	1 gm.

B. Methylene Blue (Loeffler).

Procedure

1. Fix tissues in Zenker's Fluid for twenty-four hours; wash in running water for two or three hours; dehydrate; clear; embed in paraffin wax in the usual manner.
2. Sections, 4 to 5 μ in thickness, are stained from ten to thirty minutes in Solution A; then washed with distilled water.
3. Stain with Methylene Blue (Loffier) for fifteen to sixty seconds; then wash with water.
4. Dehydrate and differentiate for a few seconds in absolute alcohol; then clear in xylol and mount.

Results

Negri bodies are stained crimson against a blue background.

Note. The method is stated to be excellent for Borrel bodies also.

Detection of Sex Chromatin

Cresyl Fast Violet Method

Solution required

A. Absolute alcohol	equal	
Ether	volumes	
B. Cresyl fast violet CNS		1 gm.
Alcohol, 50%		100 ml.

Procedure

1. Take scrapings, from the inside of the check, with a spatula.
2. Spread the scraping on a slide.
3. Fix immediately in the alcohol—ether mixture, for at least fifteen minutes.

   Note. The author (Moore) states that the usual causes of poor smears are (a) not scraping firmly enough, with the result that the number of cells obtained is insufficient, and (b) allowing the smear to dry before fixation, resulting in poor nuclear detail.
4. Immerse the preparation in 70% alcohol for two minutes.
5. Immerse in 50% alcohol for two minutes.
6. Wash for one minute in each of two changes of distilled water.
7. Stain in the cresyl fast violet solution for seven to eight minutes.
8. Differentiate by dipping the slide quickly about five to seven times in 95% alcohol.
9. Dehydrate in absolute alcohol for one minute.
10. Mount in clearmount or michrome mountant, or proceed as follows:
11. Wash in xylol.
12. Mount in clearmount or michrome mountant or D.P.X. or, if these are not available, in neutral Canada balsam in xylol, although this is liable to cause fading of the stain.
13. Examine under the oil immersion objective with a strong source of light.

Results

The sex chromatin is visible only in the nucleus from the female cells.

15

HISTOLOGICAL STAINS

Many compounds are coloured but not all of them are stains. A stain or dye is a coloured compound that can be bounded by a substrate. In histological staining dyes are used to impart colours to the various components of tissues. Sometimes the colour procedure has a high degree of chemical specificity, so that a stain can be used as a histochemical reagent.

There are three general methods of staining the first of these is "direct staining", in which soaking the objects or sections in the stain is sufficient to colour the desired structure. This is rarely employed and can be used only with those stains which are entirely specific for the structure which is to be emphasized. It is far more usual to employ a process of "indirect staining" in which the whole object or section is first uniformly stained in the dye selected and the "differentiated" in some solution which removes the dye from all those structures other than the once which are to be emphasized. The third method of staining is "mordant staining" in which some preliminary solution is used which causes the stain to "bite" into a structure without being obsorbed by others. These three great categories are by no means sharpely divided because many of the so-called direct stains actually incorporate mordants, and many of the indirect staining solutions may be used by direct method if they are first greatly diluted with water.

Before going for details of staining, one must have the knowledge of the effects of fixative on stains.

Effects of Fixative on Stains

It is not so much the chemicals of the fixatives themselves that affect the later staining of various materials as the chemical nature of the various tissues plus the compounds resulting from chemical reaction involving these tissues and the reagents in the killing and fixing fluids which, to a great extent determine the results that are obtained with different dyes.

Let us take some examples:

(a) Safranin has a staining affinity for chromatin Fluid containing chromic acid and formatin, perhaps by a sort of mordanting action, leave chromatin containing structures such as nuclei and chromosomes in a nearly perfect condition for staining with safranin. If, however, chromic acid reacts with substance in the cells to leave them coloured dark brown, the safranin is correspondingly dulled, the dullness and brownness increasing with the intensity of the reaction between the acid and the cell contents. A point will be reached where there is no longer any differentiation between cytoplasm and other cell contents (as when the tissue are saturated with tannin compounds) and bleaching as a prelude to staining becomes necessary.

(b) Some killing fluids pressure the plastin of mitotic figures perfectly, of these fluids some mordant the plastin so that if stains but other fails to mordant it. This explains why certain staining procedures which are supposed to reveal the so-called "Spindle fibres" occasionally fail, the fault is not because the staining was not carried out perfectly but because the structures involved were inadequately mordanted by the preservatives. Navashin's fluid, for example, preserves plastin but rarely mordants it.

(c) The triple combinations work as they should only on material originally fixed in a fluid that contains osmic acid or on sections that were mordanted in a weak solutions of osmic acid in chromic acid. Otherwise, the chromosomes tend to appear purple rather than red, i.e., they have an affinity for violet rather than for safranin.

Bleaching

Sections cut from materials which were exclusively darkened by the fixing fluid require to be bleached before a proper stain can be obtained. One should not, however, bleach sections without being fairly certain that it should be done because bleaching frequently is more or less damaging the tissues. Many types of material which are naturally dark (such as the mechanical tissue in fern rhizome),. do not actually require bleaching and can be beautifully stained if a proper selection of dye is made

(a) If the section appear to have suffered from too much chromate fixation, immerse the section far one minute in a 1% aqueous solution of potassium permanganate, rinse, remove the permanganate in a 1% aqueous solution of oxalic acid, wash again and proceed to the staining.

(b) Chlorine gas fumes bleach excellently. Place enough crystals of potassium chlorate to approximate the size of a grain of wheat in coplin and pour a little diluted hydrochloric acid. As soon as the greenish-yellow fumes starts coming, fill the jar quickly with 50% alcohol. About 20 minute minimum immersion of the sections is required to effect bleaching. Sections, however, cannot be stained if commercial bleaching fluids containing chlorine are used.

(c) Boric acid solution has been used for bleaching as has a 2% aqueous solution of ammonium persulphate. The latter should be used with caution as it is a powerful oxidizer.

To restore the staining capacity of tissues bleach if necessary, then soak for fifteen minutes in a 10% solution of benzol peroxide in acetate, wash out in a solution of 2 parts xylol to 3 parts of acetate followed by absolute alcohol.

Differential Acidification

In order to identify and localize stains in tissues whose elements do not present sufficient contrast in their relative degree of acidic or basic reaction, the hydrolysis phase of Feulgen's reaction may be employed to render nuclei more acid more and thus to give them a greater affinity for nuclear stains.

The procedure is especially adaptable to more or less thick materials which are to be mounted entire—such as fern, prothallia, leaves and similar other objects. The slide or fixed materials are brought down to water, placed in cold 1/N HCI for exactly one hour, rinsed with one change of distilled water and then placed in the nuclear stain. The acid must not be heated as is done in the regular Feulgen technique, otherwise the material is very likely to become dissociated.

Progressive and Regressive Staining

By observing the process of the staining under the microscope from time to time, any desired intensity may be attained. This is known as "progressive staining". A sharp differentiation usually cannot be obtained by this method. Consequently, the general practice is to overstain considerably and then to destain or otherwise to differentiate until a satisfactory optimum has been reached. This is known as "regressive staining".

General Staining

A general stain is one that stains everything indiscriminately. There is little or no selective differentiation, but such may to a slight extent be obtained by regressive destaining.

Classification of Dyes used in Microscopy

With few exceptions, notable haematoxylin and carmine, the stains used for colouring microscopic tissue preparations are synthetic dyes. These dyes constitute a broad class of aromatic organic chemicals which can be divided and subdivided into various groups according to their chemical and physical properties. Brief descriptions of some of these divisions are given below:

Acid Dyes

Strictly speaking, dyes of this class are those in which the balance of the charge on the dye-ion is negative. Acid dyes, with some exceptions, stain tissue elements that are basic in reaction.

Basic Dyes

A dye of this class is one in which the charge on the dye-ion is positive, but not all so-called dyes are ionic. Basic dyes stain tissue elements that are acidic in reaction, but there are also exceptions.

Amphoteric Dyes

Under certain conditions a number of dyes are amphoteric. This occurs among both acid and basic dyes. Many acid dyes possess both reactive acidic groups and reactive basic groups, in addition to their cations. These groups together confer amphoteric properties on the particular dye. Such dyes are normally regarded as being acid dyes. With one exception, there are no basic dyes whose molecules possess both reactive basic and reactive acidic groups, in addition to their anions, the exception being nono-fuchsinic acid, recently discovered. This somewhat bizarre compound might be regarded as a strongly acid basic dye.

Neutral Dyes

Most of these are not really dyes at all, but are non-ionic organic colouring matters. Some of them contain acidic or basic groups, however, and for this reason such neutral "dyes" are sometimes classified as "acid" or "basic". Among the neutral "dyes" are the Sudan colours (1, 2, 3, 4 etc.). This type of "dye" is probably absorbed or dissolved by lipid to which colour is imparted. A few non-ionic "dyes" such as **Sudan Black**, which has two basic groups (inino), might conceivably unite chemically with acidic elements of certain tissues. All "neutral dyes" can be converted into true dyes by the process of sulphonation. They then become water-soluble dyes. In the strict sense a dye is a coloured aromatic, organic substance capable of electrolytic dissociation which enables it to unite by chemical and/or physical means with tissue-elements or other substances of opposite charge.

Neutral dyes should not be confused with "neutral" or compound stains. The latter are formed by chemical union between a basic and an acidic dye. Such compounds of basic and acid dyes have long been known to histologists. It is doubtful, however, if the average textile dyestuff manufacturer is aware of such preparations. It is still more doubtful if they would be of any interest to textile dyer.

Compound Dyes

As stated above these are neutral dyes of the kind that appear to be known only to the histologist. They are produced by neutralizing suitable acid dyes with suitable basic dyes. That is to say, the negatively charged dye-ion of the acid dye units with the positively charged dye-ion of the basic dye when aqueous solutions of the two dyes are brought together, to form a precipitate which is a two-colour dye. If, however, one of the dyes is a polychronic stain, then the resultant compound stain will also possess the same polychrome properties plus the colour of the other dye. Such a compound dye is a true chemical compound and not a mixture. Among such dyes are Jenner stain *Jenner* (1899), Leishman stain *Leishman* (1901), methyl green organge *G. Kardos* (1911, neutral red-light green *Twort* (1924) and neutral red-jast green FCF *Ollet* (1947, 1951).

Another series of compound dyes known to histologists consists of a mixtures of acid dyes of the Masson types. But these are not chemical compounds. Then there are mixtures of basic dyes such as methyl green-Pyronin. These are mixtures, not chemical compound.

Although it is possible to form chemical compound dyes by the union of many basic dyes with many acid dyes ; as well as by the interaction of certain acid dyes with certain other acid dyes, it is not possible to produce chemical compound dyes by the interaction of any basic dyes with any other basic dyes. The reason for this is that there are no basic dyes, as there are acid dyes, whose molecules contain both reactive basic and acidic groups on the dye-ion itself.

Mordant Dyes

By colour index (convention) the mordant dyes are defined as those used in conjunction with metal salts. The mordant may be applied before the dye, together with the dye as a soluble dye-metal complex or after the dye. In some instances a mordant dye is also an anionic or, rarely, a cationic dye in its own right. Mordant dyes have many uses in histology and histochemistry.

Some of the important and frequently used stains are given below:

Haematoxylin and Eosin

The method described here is typical of many H and E procedures.

Solution required

A. *Mayer's haemalum*

Dissolve the following, in the order given, in 750 ml of water.

Aluminium potassium sulphate	— 50 gm
Haematoxylin (C.I. 75290)	— 1.0 gm
Sodium iodate	— 0.1 gm
Citric acid (monohydrate)	— 0.1 gm
Chloral hydrate	— 50 gm

Make up with water to 1000 ml. The solution may be reused many times. If the solution fails to stain nuclei property, it may be over-oxidized.

B. *Eosin*

Eosin (C.I. 45380)	— 2.5 gm
Water	— 500 ml.

This solution keeps indefinitely and may be used repeatedly. Moulds often grow in it and need to be removed by filtration. Alternately, use 0.2% ethyl eosin (C.I. 45286) in 95% alcohol.

Procedure

1. De-wax and hydrate paraffin sections. Frozen sections should be dried onto slides.
2. Stain in Mayer's haemalum (solution A) for 1-15 minutes (usually 2-5 mts). Overstained sections can easily be differentiated by agitating for a few seconds in 1% (v/v) concentrated hydrochloric acid in 95% alcohol, then washing thoroughly in tap water.
3. Wash in running tap water for 2 or 8 mts or until the sections turn blue. If the tap water is not sufficiently alkaline to blue the sections, add a few drops of ammonium hydroxide or of saturated aqueous lithium carbohydrate or a small pinch of calcium hydroxide to about 500 ml or water and leave the washed sections in this for 30-60 sec, then rinse in tap water again. Examine the wet slide under a microscope to check that selective nuclear staining

has been achieved. Any blue colouration of cytoplasm and connective tissue should be extremely paint.

4. Immerse slides in eosin soln for 30 sec. With agitation.
5. Wash and differentiate in running tap water for about 30 sec.
6. Dehydrate in 70%, 95% and two changes of absolute ethanol (with agitation, about 30 sec. in each change; without agitation 2-3 mts in each change).
7. Clear in xylene and cover, using a resinous medium.

Results

Nuclear chromation—blue or purple, cytoplasm, collagen, keratin, erythrocytes-pink.

Ehrlich's Haematoxylin

Solution required

Haematoxylin (Ehrlich)
Potassium permanganate — 0.1%

Procedure

1. Material should be fixed in 10% formalin and embedded in paraffin wax.
2. Fix sections to slides and bring down to distilled water as usual.
3. Stain in Ehrlich haematoxylin for 10 mts.
4. Pour off excess stain, then rinse and blue in tap water.
5. Immerse in potassium permanganate 0.1% for 10 sec., then wash well with water.
6. Dehydrate; clear; mount in balsam.

Results

Keratohyalin is stained blue-back while the other elements are unstained or faintly stained.

Mallory's Phosphotungstic Acid Haematoxylin

Solution required

Haematoxylin 10% in absolute alcohol (ripened for 3 months or more)	— 1 ml
Phosphotungstic acid	— 2 gm
Distilled water	— 100 ml

Procedure

1. Fix in Zenker Embed in paraffin wax.
2. Bring sections down to distilled water.

3. Treat with iodine to remove mercuric precipitate.
4. Remove iodine with 0.5% aqueous sodium hyposulphite.
5. Wash thoroughly in running water.
6. Immerse for five to ten minutes in 0.25% potassium permanganate; then wash in tap water.
7. Immerse for ten to twenty minutes in 5% oxalic acid; then wash thoroughly with tap water.
8. Stain twelve to twenty-four hours in haematoxylin solution prepared as above.
9. Wash in tap water; dehydrate with 95% and absolute alcohol.
10. Clear in xylol and mount.

Results

Nuclei, centrioles, achromatic spindles, fibroglia, myoglia, fibrin, contractile elements of straited muscle—turn blue. Collagen, reticulum, ground substances of cartilage and bone—turn yellowish to brownish red. Coarse elastic fibrils turn faint purple.

Harris' Haematoxylin

Solution required

Haematoxylin crystals	— 2 gm
Aluminium chloride	— 1 gm
Alcohol 50%	— 1000 ml

Heat on a waterbath until dissolved. Add 6 gm mercuric oxide and fitter. Add to the filtrate 1 ml conc. HCI.

Procedure

1. Bring slides down to water, or wash the killing fluid out of the material with water.
2. Rinse for about 20 minutes in the haematoxylin solution.
3. Rinse in distilled water until all the excess stain has been washed away.
4. Destain section in acid water (about 5 drops or less of HCl to each 100 cc of water). As a rule the time is about 5 sec., but one can easily stop the action of the acid, rinse the material with tap water (which will blue the stain).
5. Wash for a few seconds in tap water. If water is not sufficiently alkaline to blue the materials, and 1 or 2 drops of ammonia to the water. If it is desired to clear the cytoplasm, place the slide

or materials in a weak solution of lithium carbonate in water. About 10 mts suffices.

6. Rinse in water again, then proceed with the dehydration and mount as usual.

Result

Useful stain for plant materials such a fern prothallus, the Rhodophyta, many fungi and similar objects.

Delafield's Haematoxylin

Solution required

Haematoxylin crystals	— 1 gm
Absolute alcohol	— 6 ml
Ammonia alum	— 100 ml
Glycerine	— 2.5 ml
Methyle alcohol	— 2.5 ml

Procedure

1. Dissolve haematoxylin crystals in absolute. Add this drop to 100 ml of saturated solution of ammonia alum. Expose to light in open bottle for one week. Filter and add glycerine and methyl alcohol. Pour into a shallow dish and expose to a quartz mercury vapour lamp for 2 hrs.
2. Transfer the material to the stain fern either water or 50% alcohol. The length of time for which the sections may be left in stain depends parsy on the character of the material, partly on the nature of the killing fluid. Roughly 3-30 mts is time taken.
3. Wash in running tap water a few minutes to remove completely all excess stains. The washing should be through to avoid the precipitation.
4. Treat with aciduate water, and then transfer quickly to water and wash in running tap water until the section acquire a rich purple colour.
5. Pass through 50%, 70% and 95% alcohol.
6. Rinse in xylol and mount in balsam.

Heidenhain Iron Haematoxylin

Solution required

(A) Iron alum	— 4 gm
Distilled water	— 100 ml

Dissolve by shaking. Do not heat.

(B) Haematoxylin 10% in absolute
Alcohol (ripened for 3-6 months) — 5 ml
Distilled water — 95 ml

Procedure

1. Mordant sections in A for ½ to 24 hrs, the time varies with material and fixative.
2. Rinse with water.
3. Transfer to solution B and allow to remain the same time as in stage 1. The sections should not be an uniform pitch black and show no details.
4. Rinse in water.
5. Differentiate in solution A; control destaining under the microscope.
6. Wash in running water for 5 mts.
7. Counter stain if desired.
8. Dehydrate, clear and mount.

16

HISTOCHEMICAL STAINS

Histrochemical techniques enable the identification and localization of specific substances within tissues. The methods depend on chemical reactions between the substance to be identified and localized in a tissue section, and one or more reagents in which the tissue section is incubated. The histochemist tries to rearrange matters so that the end product of the chemical reaction is both coloured and insoluble and therefore easily visible on microscopy.

Now the histochemists are equipped with adequate tools for further studies and therfore the 1 number of substances for which histochemcial methods have been devised is increasing rapidly. Adequate histochemical methods now exist for the demonstration of carbohydrates, proteins, lipids, nucleic acids, enzymes etc. At present, in this chapter we consider the staining techniques for carbohydrates proteins and lipids. Rest are discussed elsewhere in this book.

CARBOHYDRATES

Although the term "carbohydrates" embraces all the sugars and their derivatives, the only such substances available in sections of fixed tissue are those in which the sugars form parts of macromolecules. The compounds contain mainly carbon, hydrogen and oxygen. It is not possible to demonstrate many of the individual carbohydrates at the present time because of their solubility.

GLYCOGEN

This is the only member of the polysaccharide. Group which can normally be demonstrated histochemically in tissue sections. Glycogen, which is polymerized glucose, is a normal piolysaccharide it is a storage carbohydrate and accumulate in the hepatocytes in the liver. It is also demonstrable in skeletal and cardiac muscle, as well as in many organs or the new born. Glycogen will rapidly break down to glucose after death, so it is imperative to obtain rapid and adequate fixation. A great deal has been written about the fixation of glycogen, especially in regard to its possible solubility in water. Conflicting opinions are still expressed about the choice of fixative weather to use alcoholic fixatives, Picric acid fixatives or any standard solution. Lillie (1947) showed that little or no loss of glycogen was observed after normal saline fixation, followed by a lengthy wash in running tap water. Personally I prefer to use Gendre's fixative at 4°C on thin slices of tissue. Murcury containng fixatives should be avoided.

Glycogen can be routinely demonstrated by Best's carmine Schiff methods silver techniques or by iodine. The silver and iodine methods are the least specific and are little used today. The best's carmine method can prove capricious and a known positive control should always be used.

Best's Carmine Method

Reagents required

1. Carmine
2. Potassium carbonate
3. Potassium chloride
4. Ammonia (880)
5. Methyl alcohol
6. Absolute alcohol
7. Distilled water

Preparation of solutions

1. Best's carmine stock solution

Carmine	2 g
Potassium carbonate	1 g
Potassium chloride	5 g
Distilled water	60 ml

Boil the solution gently for 5 min allow to cool then filter. To the filtrate add 20 ml of ammonia (.880).

2. Best's carmine staining solution

Stock solution	12 ml
Ammonia (.880)	18 ml
Methyl alcohol	18 ml

3. Best's differentiator

   Absolute alcohol 8 ml

   Methly alcohol 4 ml

   Distilled water 10 ml

Sections

Paraffins sections

Freeze dried sections (recommended)

Frozen sections

Procedure

1. Bring sections from xylene to 70 percent alcohol
2. Place sections in 1 per cent celloidin 5 min
3. Transfer sections to tap water
4. Stain in alum haematoxylin 10-15 min
5. Wash briefly in tap water
6. Stain in solution (2) (Best's carmine staining solution) 30 min
7. No water, rinse sections in 2 changes of solution (3)
8. Wash sections in 90 per cent alcohol briefly 20,
9. Placė in absolute alcohol
10. Transfer to xylene
11. Mount in DPX.

Result

Glycogen : red

Nuclei : blue.

Bauer-Feuglen Method

Reagent required

1. Chromium trioxide
2. Schiff's reagent
3. Sodium metabisulphite
4. Hydrochloric acid
5. Distilled water

Staining Solutions

1. Oxiding solution (4 percent chromic acid)

   Chromium trioxide 4 g

   Distilled water 100 ml

2. Schiff's reagent

3. Sulphurous acid rinse

10 per cent sodium metabisulphite	5.0 ml
Hydrochloric acid	5.0 ml
Distilled water	90 ml

Sections

This method is recommended for the demonstration of glycogen in frozen sections (crystal, etc.).

Procedure

1. Bring all sections to water	
2. Place in solution (1) (chromic acid)	30 min
3. Wash well in tap water	10 min
4. Place sections in Schiff's reagent (solution 2)	15 min
5. Transfer to solutions (3) (sulphurous acid rinse)	2 min
6. Transfer to fresh solution (3) (sulphurous acid rinse)	2 min
7. Transfer to fresh solution (3) (sulphurous acid rinse)	2 min
8. Wash in tap water	5 min
9. Counterstain in haematoxylin	2-5 min
10. Wash in tap water	5 min
11. Dehydrate through graded alcohols to xylene	
12. Mount in DPX	

Results

Glycogen : red

Nuclei : blue

Remarks

This technique gives the best result for demonstrating glycogen in frozen sections other carbohydrates substance over oxidized and are not stained.

Preparation of Schiff's Reagents

1. *de Tomasi* (1936). Dissolve 1 g of basic fuchsin in 200 ml of distilled water and boil. Shake the solution for 5 min and allow to cool. When the temperature is down to 50°C filter and to the filtrate add 20 ml N-hydrochloric acid. Cool to 25°C and add 1 g of sodium metabisulphite. Store the solution in the dark, overnight. To this solution, add 2 g of activated charcoal and shake for 1 min. Filter and store the filtrate in a dark bottle at 4°C

Note. Always allow the aliquot of solution to reach room temperature before use and discord it after use to avoid contamination of stock solution.

2. *Barger and Oe Lamater* (1948). Dissolve 1 g of basic fuchsin 400 ml of boiling distilled water, cool to 50°C and filter. To the filtrate add 1 ml of thioryl chloride. Stand in the dark overnight. Add 2 g of activated charcoal, shake well and filter. Store the filtrate at 4°C in a dark bottle.

   Note. Allow the aliquat of Schiff's solution to reach room temperature before use and discard it after use.

PAS Reaction

Reagents required

1. Periodic acid
2. Basic fuchsim
3. Hydrochloric acid
4. Sodium metabisulphite
5. Activated charcoal
6. Distilled water

Solution

1. Periodic acid 1 per cent

Periodic acid	1 g
Distilled water	100 ml

2. Schiff's reagent (de Tomasi)

Sections

All types.

Procedure

1. Bring all sections to water	
2. Place sections in (periodic acid) solution (1)	5-8 min
3. Wash in tap water	3 min
4. Wash in distilled water	1 min
5. Treat with Schiff's reagent (solution 2)	15 min
6. Wash in tap water	10 min
7. Counterstain in haematoxylin	
8. Wash in tap water	5 min
9. Differentiate if necessarary in 1 per cent acid alcohol	
10. Wash in tap water	5 min

11. Dehydrate through graded alcohols to xylene
12. Mount in DPX

Result

PAS-Positive material :magenta

Nuclei : blue

Acid Mucosubstances : Alcian Blue Modified

Reagents required

1. Alcian Blue
2. 3-per cent acetic acid
3. 10 per cent sulphuric acid
4. Mayer's carmalum

Staining solutions

Alcian Blue pH 2.5

Alcian Blue	1 g
3 per cent acetic acid	100 ml

Alcian Blue pH 0.2

Alcian Blue	1 g
10 per cent sulphuric acid	

Sections

All types, freeze dried reommended.

Producer

1. Bring all sections to water
2. Strain in Alcian blue solution of choice — 5 min
3. Wash briefly in distilled water
4. Counterstain in Mayer's carmalum — 2 min
5. Wash in tap water
6. Dehydrate through graded alcohols to xylene
7. Mount in DPX

Results

Alcian Blue pH 0.2 : strongly sulphated acid mucosubstances blue.

Alcian Blue pH 2.5 : most acid mucosubstances blue.

Reamrks

1. Alcian Blue stains should be filtered before use
2. Counterstaining should always be light
3. For critical staining see.

Alcian Blue—CEC Method

Reagent required

1. Alcian Blue
2. Acetate buffer Ph 5.8
3. Magnesium chloride

Stock Alcian Blue stain

Alcian Blue	50 mg
Acetate buffer pH 5.8	100 ml

This stock solution is used with the necessary amount of magnesium chloride. To obatin the following molarities the stated figure Alcian Blue solution.

0.6 M=1.2 g, 0.3 M=6.1 g, 0.5 M=10.15 g, 0.7 M=14.2 g, 0.9 M =18.3 g.

Sections

All types.

Procedure

1. Bring all sections water
2. Stain in Alcian Blue solution for at least 4 hours, overnight preferably.
3. Rinse in distilled water.
4. Dehydrate through graded alcohols to xylene
5. Mount in DPX

Results

Positive with 0.06 M carboxyl and sulphated mucosubstances.
Positive with 0.3 M weakly and strongly sulphated muco-substances.
Positive with 0.5 M strongly sulphated mucosubstances.
Positive with 0.7 M highly sulphated connective tissue mucins.
Positive with 0.9 M keratan sulphate only.

Remarks

1. Background staining can occur with the low molarity solutions.
2. For comments of specificity of method.
3. Counterstain is optional.

Alcian Blue-PAS Method

Reagents required

1. Alcian Blue solution

2. 1 per cent periodic acid
3. Schiff's reagents

Staining solutions

1. Alcian Blue pH 2.5
2. Schiff's reagents.

Procedure

1. Bring all sections to water
2. Stain Alcian Blue solution **5 min**
3. Wash in distilled water
4. Oxidize in 1 per cent periodic acid **5 min**
5. Wash well in distilled water
6. Place in Schiff's reagent **8 min**
7. Wash well in running tap water **10 min**
8. Counterstain lightly in Mayer's haematoxylin if required
9. Differentiate in 1 per cent acid alcohol, blue etc
10. Dehydrate through graded alcohols to xylene.
11. Mount in DPX.

Results

Acid mucosubstance : blue

Neutral mucosubstance : red

Mixtures : purple

Remarks

1. A good method to separate acid and neutral mucosubstance
2. As Cook (1974) states, the method if negative can be taken to mean that a given susbstance is unlikely to be a mucosubstances.

Acid Mucosubstances : Dislysed Iron Method (Modified by Muller and Mowry)

Reagent required

1. Ferric chloride
2. Acetic acid
3. Hydrochloric acid
4. Potassium ferrocyanide
5. Distilled water.

Preparation of solutions

1. Stock colloidal iron solution

29 per cent ferric chloride 2.2 ml

Distilled water 12 ml

Pour the ferric chloride into boiling distilled water and stir well. The solution will turn dark red; at this point it is removed from the heat and allowed to cool. It is important that the solution is boiling when the ferric chloride is added.

2. Staining Solutions of Colloidal iron

   Glacial acetic acid 5 ml

   Distilled water 15 ml

   Stock colloidal iron solution 20 ml

3. Acid Ferrocyanide Mixture

   Potassium ferrocyanide 2 g

   Hydrochloric acid (conc.) 2 ml

   Distilled water 98 ml

   The potassium ferrocyanide is dissobled in the distilled water and the hydrocholoric acid is added to this solution.

Sections

All types.

Procedure

1. Bring all sections to water
2. Rinse in 12 per cent acetic acid.
3. Stain in solution (2), colloidal iron solution 1 h
4. Rinse in 12 per cent acetic acid, 4 charges, 3 min each
5. Treat section with acid ferrocy anide, solution (3) 20 min
6. Wash in distilled water
7. Wash in tap water 5 min
8. Counterstain in Mayer's carmalum 10 min
9. Wash in tap water
10. Dehydrate through graded alcohols
11. To xylene and mount in DPX

Results

Acid mucosubstances : bright blue

Nuclei : red

Remarks

Some workers prefer to use the Feulgen reaction as counterstain.

Acid Mucosubstance : Toluidine Blue Method

Reagents required

1. Toluidine Blue
2. Absolute Alcohol
3. Distilled water

Staining solution

Toluidine Blue	100 mg
Absolute alcohol	30 ml
Distilled water	70 ml

PROTEINS

Proteins are the third of the components that go to make up cell cytoplasm the other two being lipids and carbohydrates. Proteins are tc be found in all cells and tissues. The actual demonstration of protein material is nor difficult, but the identification of any particular protein is difficult. The simple proteins are the naturally occurring proteins that upon hydrolysis will yield amino acids only. The conjugated proteins consist of simple protiens combined with a non-proteins material, for example lipid in lipoproteins and nucleic acids in nucleoproteins.

The methods discussed below do not demonstrate the protein molecule as a whole, but are dependent on the presence of certain types of reactive groups or special types of linkages within the protein molecule. Histochemical reactions are available for some, but not all of the reactive groups of the protein molecules . Some of the important methods are discussed as follows.

The Ninhydrin-Schiff Method

Reagents required

1. Ninhydrin
2. Absolute alcohol
3. Schiff's reagent

Preparation of solutions

1. Ninhydrin 0.5 per cent solution

Ninhydrin	500 mg
Absolute alcohol	100 mg

2. Schiff's reagent

Sections

Freeze dried

Paraffin sections

Cryostat unfixed

Cryostat prefixed

Procedures

1. Bring sections down to 70 per cent alcohol
2. Treat with solution (1), ninhydrin, at 37°C overnight
3. Wash in running tap water
4. Place sections in Schiff's reagent solution (2), for 45 min
5. Wash well in running tap water
6. Counterstain if required in haematoxylin
7. Wash in tap water
8. Dehydrate through graded alcohols to xylenes and mount.

Results

α-Amino groups : pink to red

Remarks

1. 1 per cent Alloxan, also in absolute alcohol, may be used instead of 0-5 per cent ninhydrin.
2. Control sections may be needed for other PAS-positive material.

DNFB Method for Tyrosine, SH and NH_2 Groups

Reagents required

1. 2-4 Dinitrofluorobezene
2. Ethyl alcohol
3. Sodium hydrogen carbonate
4. Sodium hydrosulphite
5. Sodium nitrite
6. Hydrochloric acid
7. H-acid (8 amino-1-naphthol 3, 6-disulphonic acid)
8. Veronal acetate buffer pH 9.4

Preparation of solutions

1. *Dinitrofluorobenzene solution*

   2-4 dinitrofluorobenzene saturated in 90 per cent ethyl alcohol saturated with sodium hdrogen carbonate.

2. *Nitrous acid*

5 per cent sodium nitrite	**10 ml**
2_N hydrochloric acid	**40 ml**

3. *'H'-acid*

   H-acid saturated in 0.1 m veronal acetate buffer pH 9.4

Procedure

1. Bring section to absoulte alcohol and allow to dry in air
2. Place in DNFB solutions 2-16 h (overnight is usally ideal)
3. Rinse in 90 per cent alcohol (3 changes) then tap water
4. Treat with 5 per cent sodium hydrosulphite at 45°C for 30 min
5. Wash in distilled water
6. Treat sections with nitrous acid solution at 4°C for 30 min
7. Wash in cold distilled water
8. Treat with H-acid solution at 4°C for 15 min
9. Wash in tap water.
10. Dehydrate through graded alcohols to xylene
11. Mount in DPX

Results

A positive reaction is reddish purple

Sakaguchi Method Modified by Baker

Reagents required

1. Sodium hypochlorite
2. α-Naphthol
3. Sodium hydroxide
4. Pyridine
5. Chloroform
6. 70 per cent Alcohol

Preparation of solutions

Solution 1

1 per cent Sodium hydroxide

Solution 2

α-Naphthol	1 g
70 per cent Alcohol	100 ml

Solution 3

Milton	1 ml
Distilled water	99 ml

Incubating solution 4

Solution (1)	2 ml
Solution (2)	2 drops
Solution (3)	4 drops

Pyridine-chloroform solution 5

Pyridine	30 ml
Chloroform	10 ml

Sections

Freeze dried
Cryostat
Paraffin sections

Suitable control sections

Testis

Procedure

1. Bring all sections to water
2. Rinse in 70 per cent alcohol
3. Cover section with incubating solution (4), 15 min
4. Drain and blot dry
5. Immerse in pyridine-chloroform, solution (5), 2 min
6. Mount in pyridine-chloroform mixture and ring coverslip

Result

Arginine : Orange-red

Remarks

The slide should be looked at microscopically immediately.

DMAB-nitrite Method

Reagents required

1. *p*-Dimethylaminobenzaldehyde
2. Hydrochloric acid
3. Sodium nitrite
4. Acid alcohol (1 per cent)

Prepartion of solutions

1. *p-Dimethylaminobenzaldehyde*

p-Dimethylaminobenzaldehyde	5 g
Hydrochloric acid (conc.)	100 ml

2. *Sodium nitrite*

Sodium nitrite	1 g
Hydrochloric acid (conc.)	100 ml

3. *Acid alcohol*

Hydrohloric acid (conc.)	1 ml
70 per cent Alcohol	99 ml

Sections

Freeze dried

Paraffin sections

Cryostat

Suitable control sections

Pancreas, duodenum, pituitary

Procedure

1. Bring sections to alcohol
2. Celloidinize in 0.5 per cent celloidin
3. Place sections in solution (1), DMAB 1 min
4. Transfer sections to solution (2), sodium nitrite 1 min
5. Wash carefully in tap water 30 sec
6. Rinse sections in solution (3), acid alcohol 15 sec
7. Dehydrate through graded alcohols to xylene and mount

Results

Tryptophan : deep blue

Remarks

This method gives good localization

Million reaction (1849), Baker (1956)

Reagent required

1. Mercuric sulphate
2. Sulphuric acid
3. Sodium nitrite

Preparation of solutions

Solution 1

Distilled water	90 ml
Sulphuric acid (conc.)	10 ml

To this solution add 10g mercuric sulphate and hear until dissolved. Cool to room temperature and add 100 ml distilled water.

Solution 2

Sodium nitrite	250 mg
Distilled water	10 ml

Staining solution 3

Solution 1	10 ml
Solution 2	1 ml

Sections

Freeze dried
Paraffin sections
Cryostat sections
Celloidin sections

Suitable controls

Pancreas, duodenum

Procedure

1. Bring all sections to water
2. Place sections in a small beaker,add solution (3) and boil gently — 2 min
3. Allow to cool to room temperature
4. Wash sections in distilled water — 2 min
5. Repeat wash in distilled water — 2 min
6. Repeat wash in distilled water — 2 min
7. Dehydrate through graded alcohols to xylene and mount in DPX

Result

Tryosine : red pink or yellowish red.

Diazotization Coupling Method for Tyrosine (Glenner and Lillie 1959)

Reagents required

1. Sodium nitrite
2. Acetic acid
3. 8-Amino -1- napthol -5-sulphonic acid (S-acid)
4. Potassium hydroxide
5. Ammonium sulphamate
6. Hydrochloric acid

Preparation of solutions

1. *Incubating solution*

Sodium nitrite	6.9 g
Acetic acid	5.8 ml
Distilled water	94 ml

2. 8-amino-1-naphthol-5-sulphonic acid — 1 g

Potassium hydroxide	1 g
Ammonium sulphamate	1 g
70 per cent alcohol	100 ml

Sections

All types

Suitable control sections

Pancreas

Procedure

1. Place sections in incubating medium at 4°C overnight in the dark
2. Rinse in distilled water at 4°C
3. Treat with solution (2) at 4°C for 1 h also in the dark
4. Rinse in 3 changes of 0-1 NHCL 5 min each
5. Wash in tap water 10 min
6. Dehydrate through graded alcohols to xylene
7. Mount in DPX

Results

Tyrosine containing protein : purple red

Ninhydrin Method (for Amino Groups)

Solutions required

1. Celloidin 0.5% in equal volumes of absolute alcohol and ether.
2. Ninhydrin (triketohydrindene hydrate). 0.5% aqueous

Procedure

1. Fix pieces of fresh material in 10% formalin for two to four hours.
2. Wash in water and cut frozen sections.
3. Wash in 70% alcohol, followed by 90% and absolute alcohol.
4. Coat sections on slides with the celloiding solution, then allow the solvent to evaporate for a few minutes.
5. Immerse sections in solution B (Ninhydrin) for one minute at 90°C.
6. Wash in water.
7. Mount in Aquamount or glycerine or glycerine jelly.

Results

Alpha amino acids and proteins : blue to violet

Notes:

(a) According to Serra (1947), the reaction is given not only by all amino acids (except proline and hydroxyproline) peptides and proline, but by amines, aldehydes, sugars with free aldehyde and keto groups and by ammonia and ammonium salts.

(b) Readers are reffered to Serra's (1946) paper for more detailed information.

Lipids

The term "lipid" is applied to a chemically heterogenous group of substances which are extracted from tissues by non-polar organic solvents such as chloroform and ether. These substances vary greatly in structural but are built from a limited number of simpler molecules joined together in different ways. These components substances do not occur in large amounts in the free state in lining tissue but are present as metasolic precursors of the lipids themselves.

There is no agreement on how lipids should be classified, it is probably easiest to make the following subdivisions.

1. Simple lipids
2. Compounds lipids
3. Derived lipids

Lipids in general terms can be termed as compounds of long chain fatty acids with an alcohol.

Simple lipids are esters of both saturated and unsaturated long chain fatty acids with alcohols compound lipids contain a non-lipid group as well as long chain fatty acids and an alcohol. On the other hand, derived lipids contain. The fatty acids that are produced by hydrolysis of the simple and compound lipids. The fatty acids can be either saturated or unsaturated.

Some individual methods for staining lipid materials are given below:

Oil Red O (Lillie and Ashburn, Modified 1943)

Reagents required

1. Oil Red O
2. Triethyl phosphate
3. Distilled water

Preparation of staining solution

Oil Red O	1 g
Triethyl phosphate	60 ml
Distilled water	40 ml

The distilled water is added to the triethyl phosphate, the dye is then added, the mixture is treated to 100°C for 5 min and is stirred constantly. The mixture is filtered when hot and a gain when cool.

This solution will keep as a stock solution but must be filtered before use.

Sections

Formalin fixed frozen sections, free floating

Cryostat post-fixed preferably free floating

Procedure

1. Wash sections in distilled water
2. Place sections in 60 per cent triethyl phosphate
3. Stain sections in Oil Red O solution at 20°C for 15 min,
4. Wash sections in 60 per cent triethyl phosphate for 30 sec.
5. Wash sections in distilled water
6. Stain sections in haematoxylin for 1 min
7. Wash sections in tap water for 5 min
8. Mount in glycerin jelly

Results

Lipid material : red

Nuclei : blue

Remarks

This method can be employed at 37°C or 60°C if required.

Sudan Black B (Lison and Dagnelie, 1935)

Reagents required

1. Sudan Black B
2. Triethyl phosphate
3. Distilled water

Preparation of staining solution

Sudan Black B	1 g
Triethyl phosphate	60 ml
Distilled water	40 ml

The distilled water is added to the triethyl phosphate, the stain is added to this solution and the mixture is heated to 100°C for 5 min stirred constantly. The mixture is filtered when hot and once again immediately before use. The solution will keep well as a stock solution but must be filtered each time it is used.

Sections

Formalin-fixed sections, free-floating

Cryostat post-fixed free floating

Procedure

1. Wash sections in distilled water
2. Place sections in 60 per cent triethyl phosphate
3. Stain sections in Sudan Black B solution at 20°C for 10 min
4. Place sections in 60 per cent triethyl phosphate for 30 sec
5. Wash in distilled water
6. Stain in Mayer's Carmalum for 3 min
7. Wash indistilled water
8. Mount in glycerin jelly

Results

Lipid material, including phospholipids; black

Nuclei : red

Remarks

This technique using triethyl phosphate can be carried out 37°C or 60°C as required.

Acidic lipids : Nile blue (Smith-Dietrich, Modified by Cain, 1947)

Reagents required

1. Nile Blue
2. Acetic acid
3. Distilled water

Preparation of solution

1. *Nile Blue 1 per cent*

Nile Blue	500 mg
Distilled water	50 ml

2. *Nile Blue 0.02 per cent*

Nile Blue	10 mg
Distilled water	50 ml

3. *Differentator*

Acetic acid, conc	0.5 ml
Distilled	50 ml

Sections

Formalin-fixed frozen sections

Cryostat pre fixed

Cryostat post-fixed

Procedure

It is necessary to stain one section by the Oil Red O or Sudan Black methods. This slide is numbered 3. Two sections are required for the following technique :

1. Bring both sections to water
2. Stain sections 1 and 2 in 1 per cent Nile Blue for 5 m in at 60°C
3. Differntiate at 60°C in solution (3) for 30 s
4. Wash in tap water
5. Mount section 1 in glycerin jelly
6. Place section 2 in solution (2) (0.02 per cent Nile Blue) for 5 min at 60°C
7. Wash in tap water
8. Differentiate section 2 in solution (3) at 60°C for 30s
9. Wash in tap water
10. Mount in glycerin jelly

Results

Section 1. Any blue staining that can be compared with a positive section 3 in taken to be acidic lipid material.

Section 2. Any red staining that can be compare with a positive results in section 3 is taken to be a non-acidic lipid.

Sections 3. Control for above

Remarks

This method has caused much controversy in the literature, and its specficity is doubtful it is however well worth applying when trying to identify an unknown lipid.

Acid Haematein method (Baker, 1946)

Reagents required

1. Formalin
2. Calcium chloride
3. Pottassium dichromate
4. Haematoxylin
5. Sodium iodate
6. Glacial acetic acid
7. Sodium tetraborate (borax)
8. Potassium ferric cyanide
9. Distilled water

Preparation of solutions

1. *Fixative*

Formalin	10 ml
Calcium chloride (anhydrous)	1 g
Distilled water	90 ml

2. *Post-chroming solution*

Potassium dichromate	5 g
Calcium chloride	1 g
Distilled water	100 ml

3. *Acid Haematein solution*

Haematein	50 mg
1 per cent Sodium iodate	1 ml
Distilled water	49 ml

Heat the solution to boiling point, allow to cool and add 1 ml of glacial acetic acid.

4. *Differentiator*

Pottasium ferricyanide	250 mg
Sodium tetraborate(borax)	250 mg
Distilled water	100 ml

Sections

For this metod unfixed pieces of tissue are recommended.

Procedure

1. Place blocks in fixative, solution (1) at 22°C, for 6-12 hrs.
2. Transfer tissue (no washing) to post-chroming solution (2) at 22°C for 18 hours.
3. Transfer tissue to fresh post-chroming solution (2) at 60°C, for 24 hours.
4. Wash in running tap water for 6 hours.
5. Cut frozen sections 10 μm thick.
6. Place sections in post chroming solution (2) at 37°C, for 1 hours.
7. Wash well in distilled water for 5 minutes.
8. Stain in acid haematin, solution (3) at 60°C for 3 hours.
9. Rinse well in distilled water.
10. Transfe. section to differentiating solution (4) at 37°C for 18 hours.
11. Wash in tap water for 10 minutes.
12. Mount in glycerin jelly.

Results

Phospholipids-dark blue

Other material may be blue.

Remarkds

The method is only specific when a control is submitted to the pyridine extraction method.

PAS Reaction

Reagents required

1. Periodic acid
2. Schiff's reagent
3. Distilled water

Preparation of solutions

1. *Periodic acid*

Periodic acid	500 mg
Distilled water	100 ml

2. *Schiff's reagent*

Sections

Cryostat post-fixed

Cryostat pre-fixed

Formalin-fixed frozen sections

Freeze dried paraffin sections

Procedure

1. Bring sections down to water	
2. Transfer sections to solution (2)	5 min
3. Wash in tap water	3 min
4. Place in Schiff's reagent	20 min
5. Wash in tap water	20 min
6. Counterstain in haemalum	5 min
7. Wash in tap water	
8. Differentiate in 1 per cent acid alcohol	5 g
9. Wash in tap water	
10. Mount in glycerin jelly	

Results

Glycolipids, mucins, eic.: red

Nuclei : blue

Remarks

It is necessary to use the follwoing controls when applying this method for glycolipids.

1. Oil Red O or Sudan Black to confirm site of possible glycolipid.
2. Aldehydes to be blocked.

Note. If formalin-fixed frozen sections are used, they must be well washed in several changes of distilled water remove any free aldehydes. If cryostat sections are used, alcohol should be used as a fixative.

Perchloric Acid-Naphthoquinone reaction (Adams, 1961)

Reagents required

1. 1 : 2- Naphthoquinone-sulphonic acid
2. Ethanol
3. Perchloric acid
4. Formaldehyde
5. Distilled water

Preparation of solution

1: 2- Naphthoquinone-4-sulphonic acid	12 mg
Ethanol	6 ml
60 per cent Perchloric acid	3 ml
Conc. formaldehyde	0.3 ml
Distilled water	2.7 ml

The ethanol-perchloric acid formaldehyde water solution is prepared first and the reagent dissolved in it.

Sections

Formol saline-fixed, frozen sections, free-floating

Formol calcium-fixed frozen sections, free-floating

Suitable sections for controls

Adrenal gland

Procedure

1. Cut frozen sections and float into formalin. Leove for 7 days
2. Mount sections on slides and dry at room temperature
3. Soak sections in reagent
4. Heat sections in reagent to 60-70°C for 10 mn
5. Mount section in 60 per cent perchloric acid

Results

Cholesterol and its esters : dark blue

Remarks

1. The dark blue colour is stable for a few hours
2. During the heating, the sections should change colour from red to dark blue.

Digitonin Method (Windaus, 1910)

Reagents required

1. Ethyl alcohol
2. Digitonin
3. Distilled water

Preparation of sections

1. *Ethyl alcohol 50per cent*

Ethyl alcohol	100 ml
Distilled water	100 ml

2. *Digitonin solution*

Digitonin	500 mg
Solution (1)	100 ml

Sections

Formalin-fixed frozen sections free-floating. A control section is stained by Oil Red O method.

Procedure

1. Incubate secisn for 3 hours in solution (2) at room temperature
2. Rinse sections in solution (1)
3. Float sections onto slides
4. Mount in glycerin jelly

Results

Digitonin section : free cholesterol, birefringent.

Oil Red O section : free cholesterol, birefringent. Cholesterol ester, stained by Oil Red O.

Copper-Rubeanic Acid method (Holczinger, 1959)

Reagents required

1. Copper acetate
2. Rubeanic acid
3. Ethanol

4. Distilled water
5. Ethylenediamine tetra-acetic acid (Disodium), EDTA

Preparation of solutions

1. *0.005 per cent Copper acetate*

Copper acetate	5 mg
Distilled water	100 ml

2. *0.1 per cent EDTA*

Ethylenediamine tetra-acetic acid	50 mg
Distilled water	50 ml

3. *0.1 per cent Rubeanic acid*

Rubeanic acid	50 mg
Absolute alcohol	35 ml
Distilled water	15 ml

Dissolve the rubeanic acid in the absolute alcohol by warming slightly then add the distilled water.

Sections

Cryostat unfixed

Cryostat pre-fixed

Formalin-fixed frozen sections

Procedure

1. Place sections copper acetate solution for 3-5 hours.
2. Wash sections in EDTA solution for 10 seconds.
3. Wash sections again in EDTA solution for 10 seconds.
4. Wash sections in distilled water for 10 minutes.
5. Immerse sections in rubeanic acid solution for 30 minutes.
6. Wash section in 70% alcohol for 3 mts.
7. Wash sections in running tap water.
8. Mount sections in glycerin jelly or dehydrate throguh graded alcohol and mount in DPX.

Results

Fatty acid—greenish black.

Luxol Fast Blue Method (For gross brain sections)

Solutions required

A.	Luxol fast blue	1 gm
	Alcohol 95%	1 liter
	Acetic acid, 5% aqueous	5 ml

B. Lithium carbonate, saturated aqueous

C. Solution B	10 ml
Distilled water	1 liter

Procedure

1. Fix brains in 10% fromalin preferably by perfusion and sliced to the desired thickness as soon as they are firm enough.
2. Store the sliced brain in the fixative, for minnimum time of two weeks for human material, or in the case of laboratory animals for at least one week.
3. Wash the slices in water for at least six hours.
4. Dehydrate for one hour in each of two changes of 95% alcohol.
5. Stain in solution A for sixteen to eighteen hours at 45-55°C
6. Remove excess stain by immersing the preparation in 95% alcohol.
7. Rinse in distilled water.
8. Differentiate in several changes of solution C.
9. Refine the differentiation in a number of changes of 70% alcohol.
10. Rinse in distilled water.
11. Store in 10% formalin.

Results

Tracts of white matter stain brilliant blue, contrasting strongly with cellular areas of grey matter, stained very pale green.

Notes: The author states that absence of myclin is evident in pathological demyelinated regions of adult brain and in the non-myelinated areas of very young animals.

Sudan Black Method (J.R. Baker's Technique)

Solutions required

A. *Formaldehyde-saline*

Formalin (Formaldehyde 40%)	10 ml
Sodium chloride 10% aqueous	7 ml
Distilled water	83 ml

Note. Keep a few pieces of marble chips in the solution to maintain neutrality.

B. Formalin (Formaldehyde 40%) neutral.

Note. Keep a few pieces of marble chips in the bottle.

C. Potassium dichromate 2.5% aqueous	88 ml
Sodium chloride 10% aqueous	7 ml

Note. Keep a few pieces of marble chips in the bottle.

D. *Dichromate-fromaldehyde*

Solution B	1 volume
Solution C	19 volumes

E. Potassium dichromate 5% aqueous

F. *Gelatine for embedding*

Gelatine powder	25 gm
Water	100 ml
Sodium-p-hydroxybenzoate	0.2 gm.

Sprinkle the gelatine on to the water and leave it to soak for an hour, afterwards in an incubator maintained at 37°C until all the gelatine has dissolved, then strain through muslin while still warm.

Note. If sodium p-hydroxybenzoate, which is added to prevent the growth of moulds and bacteria is not available in the laboratory, then 0.25 to 0.5 gm of Thymol should be used instead.

G. Formalum (for hardening gelatine)

Formalin (Formaldehyde 40%)	20 ml
Potassium alum 5% aqueous	80 ml

Keep marble chips in the bottle.

Note : Both gelatine blocks and gelatine sections be preserved indefinitely in Formalum, which makes the gelatine very hard, thereby facilitating the cutting of thin sections which are non-sticky.

Important. Formalum must not be used in the acid haematein test for phospholipids (Baker 1946) as the alum would react with the haematein.

H. Sudan Black

Sudan Black	0.5 gm
Alcohol 70%	100 ml

Boil for ten minutes under a reflux condenser; then cool and filter.

I. *Carmalum* (*Mayer*).

Procedure

1. Fix a piece of tissue not more than 3 mm thick in the formaldehyde saline for an hour
2. Transfer without washing to the dichromate formaldehyde (Solution D) and leave for five hours.
3. Transfer, without washing to 5% aqueous potassium dichromate and leave for about eighteen hours.
4. Leaving the tissue in the same solution, transfer to the paraffin oven at 60°C for twenty-four hours.

5. Wash in running water for six hours.
6. Leave overnight in the melted gelatine in the oven at 37°C.
7. Cool the gelatine, preferably in a refrigerator.
8. Cut out a rectangular block containing the specimen.
9. Immerse the block overnight (or any conveniently longer time) in formalum, placing a marble chip in the capsule or tube.
10. Cut sections 8 to 10 μ on the freezing microtome.
11. Transfer a section to 70% alcohol.

    Note. It is best to transfer sections from fluid to fluid up to stage 16 in a Royal Worcester Porcelain thimble No. a-4756. size 2.
12. Transfer to the Sudan black solution and leave for 1/2-4 minutes (The best period is usually about $2^1/_2$ minutes).
13. Wash in 70% alcohol for five seconds,
14. Wash in 50% alcohol for one minute.
15. Wash in water, sinking the section gently with a camel hair brush if it floats.
16. Transfer to Carmalum for two to three minutes. (The optimum time is usually three minutes)
17. Rinse in distilled water.
18. Transfer the section to a fairly large dish, or a tongue jar of tap water, and leave for two minutes, or any conveniently longer time.
19. Wash again in another large bowl of water.
20. Transfer to a petri dish of water.
21. Float the section on to a slide.
22. Blot away excess water but do not allow the section to dry.
23. Mount in Farrants medium or in Aquamount.
24. Attach a cap to hold the coverslip to the slide : then leave overnight in the oven to harden the mounting media, before examining the preparation under the oil immersion objective.

    Note. The slide may be examined after a quarter of an hour, if desired; then returned to the oven to complete the hardening.

Results

Lipids dark blue or blue back. Cytoplasm : colourless or pale grey-blue.Chromatin : pink or red.

Note. If the results are not good, another section should be tried with variations of the staining times.

Never attempt to judge the colouring until the section is mounted and examined under the oil immersion objective.

It is recommended that the technique be learned on the intestine of the mouse, as it is scarcely possible to fail with this. Cut out a piece of empty intestine about 1 cm long and immerse in formaldehyde saline for five minutes, then open it by a longitudinal cut from one end to the other taking care not to do any unnecessary damage to the villi.

The section should be left only one minute in the Sudan black and two and a half minutes in the carmalum.

17

SPECIFIC STAINING FOR VARIOUS TISSUES

Certain specific stains are used for identifying and studying conncetive tissues such as muscle fibres, blood cells etc. The three main fibrous connective tissues are *collagen*, *reticulin* and *elastin*. These fibres have certain physical and chemical properties which enable them to be stained individually.

Collagen is a basic glycoprotein containing high proportions of glycins and proteine. *Hydroxylysine*, an animo acid is also found in collagen in small quantity. In the elctron microscope, collegen fibres present a triple helices joined end to end and side to side by hydrogen bonds by covalent bridges seem to be bundles of these fibres.

The term "reticulin includes the basement membranes of epithelia and blood capillaries. Both types contain more carbohydratess then collagen. The third type, elastic fibres are made of elastin. The elastin is a hydrophobic protein, rich in glycine, alamine and value.

The staining tecniques for connective tissues fall into three categories those based on the use of mixture of anionic dyes to give different colours to collogen and cytoplasm, the methods for reticulin and methods for elastin. Some of the individual methods for all the three types are given as below:

STAINING METHODS FOR COLLAGEN

Many techniques are available for staining with mixtures of dyes. Collagen and cytoplasm are coloured differently but other elements such as cartilage, fibrin and secretory granules aquire characteristic colours.

Iron-haemotoxylin and Van Gieson Method

Solutions required

1. Weigers iron haematoxylin

Solution A. Haemtoxylin

(C.I. 75290):	5 g
95% ethanol	500 ml

Keeps indefinitely

Solution B. Ferric chloride

($FeCl_3.6H_2O$) :	5.8 g	
Water :	495 ml	
Concentrated hydrochloric acid:	5 ml	Keeps indefinitely

Working solution

Mix equal volumes A and B. Put A in the staining jar or tank first for more rapid mixing. The mixture should be made just before using, but can be kept for about 2 weeks at 4^0C.

2. Van Gieson's solution

Acid fuchsine

(C.I. 4285) :	0.5 g	Keeps indefinitely
Saturated aqueous picric acid:		

Procedure

1. De-wax and hydrate paraffin sections.
2. Stain in working solution of Weigert's haematoxylin for 5 minutes (10 minutes if the solution is more than a few days old).
3. Wash in running tap water.
4. Stain in van. Gieson's solution, 2-5 min. The time is not critical.
5. Wash briefly, in running tap water. This also differentiates the acid fuchsine.
6. Dehydrate rapidly three changes of 100% ethanol. This step also differentiates the picric acid.
7. Clear in xylene and mount in a resinous medium.

Result

Nuclei-black or brown; collagen-red; cytoplasm (especially smooth and striated muscle), keratin and erythrocytes—yellow.

Cason's Trichrome Staining Method

Solution required

A. Wieger's iron haematoxylin (working solution).

B. Cason's trichrome solution

Water: 200 ml

Dissolve in order stated

Phosphotungstic acid	1 g	
Orange G (C.I. 16230)	2 g	Keeps for
Aniline blue WS (C.I.42755)	1 g	1-2 years
Acid fuchsin (C.I. 42685)	3 g	

Procedure

1. De-wax and hydrate sections.
2. Stain in Weiger's haematoxylin, 5 minutes.
3. Wash in running tap water, 2 minutes
4. Immerse in Cason;s trichrome solution, 5 minutes.
5. Wash in running water, 3-5 seconds.
6. Blot slides dry with filter paper.
7. Dehydrate rapidly in three changes of 100% ethanol.
8. Clear in xylene and cover, using a resinous medium.

Results

Collagen-blue; cytoplasm, muscle-red; keratin, erythrocytes-orange; nuclei-brown (but sometimes blue). The pre-staining with iron-haematoxylin makes the trichrome colours a little unpredictable with some material. If the nuclei stain with iron-haematoxylin is omitted, most nuclei are coloured red; others may be blue or unstained.

Orcein-Picro Fuchsin Method

Solution required

A. Orcein (Unna)1% in 80% alcohol 100 ml

Hydrochloric acid, cone. 1 ml

B. Picro-fuchsin (Van Gieson).

Procedure

1. Sections are mounted on slides and brought down to 70% alcohol in the usual manner. If tissues have been fixed in a fluid containing mercury, the mercurial precipitate is removed by the standard technique.
2. Immerse is orcein solution (recipe as above) for half an hour longer if necessary : then rinse in acid alcohol.
3. Rinse in 70% alcohol; then in water.
4. Stain with piero fuchsin (Van Gieson) for three to five minutes.

5. Rinse rapidly (not move than a few seconds) in water.
6. Dehydrate rapidly; clear, then mount.

Results

Collagen fibres are stained red; elastic fibres, brown; erythrocytes, epithelia, muscle, etc, yellow.

Masson's Trichrom Staining Method

Solution required

A. Iron alum 5% aqueous
B. Regaud's haematoxylin solution.
C. Picric acid, saturated in 95%

alcohol	20 ml
Alcohol 95%	10 ml

D. Ponceau fuchsin.
E. Phosphomolybdic acid 1% aqueous.
F. Aniline Blue 5% in 2% acetic acid.

Procedure

1. Fix pieces of tissue in Bouin's fluid for three days or in Regaued's fluid for one day.
2. Wash in running water; dehydrate; clear and embed in paraffin wax as usual.
3. Section 5μ in thickness are fixed to slides; de-waxed and passed through descending grades of alcohol down to distilled water in the usual manner.
4. Mordant in Solution A for five minutes at 45°C to 50°C.
5. Wash well in distilled water.
6. Stain for five minutes in Regaud's haematoxylin at 45°C to 50°C.
7. Rinse in distilled water.
8. Differentiate in picric alcohol (solution C above) controlling by examination under the microscope, while preparation is still wet.
9. Wash in running rap water for a minute or so.
10. Stain for five minutes in the Ponceau fuchsin solution.
11. Rinse in distilled water.
12. Differentiate in the phosphomolybdic acid solution for five minutes.
13. Add 5.0 ml of the acetic aniline blue (Solutio F above) to the phosphomolybdic acid on the slide and mix by rocking the slide gently. Allow this mixture to act for five minutes.
14. Pour off excess liquid and rinse in distilled water.

15. Immerse in phosphomolybdic acid leave therein for five minutes.
16. Transfer to 1% acetic acid and leave therein for five minutes.
17. Wash in distilled water.
18. Dehydrate in 95% alcohol, followed by absolute alcohol; clear in xylol; mount.

Results

Collagen, deep blue, Neuroglia fibrils, red, Nuclei black. Argentaffin granules, black or red.

Saffron-Erythrosin Method

Solution required

A. Saffron 2 gm

Distilled water 100 ml

Boil gently for an hour; allow to cool : then filter; and 1 ml of 40% formaldehyde and 1 ml of 5% tannic acid to the filtrate.

Note. Saffron solution deteriorates after a few weeks and it is best to prepare to solution in small quantities, as required.

B. Delafield or Ehtlich haematoxylin

C. Erythrosin, 1% aqueous.

Procedure

1. Fix small pieces of tissue in Bouin, Zenker-formaldehyde or in mercuric formaldehyde.
2. Wash; dehydrate; embed.
3. Sections are stained for five or ten minutes with Delafield or Ehtlich haematoxylin; rinse in water
4. Blue in tap water in the usual manner or in 1% sodium phosphate (Na_2HPO_4).
5. Stain for two to five minutes in 1% aqueous erythrosin.
6. Rinse quickly with water.
7. Differentiate with 70% alochol for few seconds, controlling under the microscope until the collagen fibres are nearly colourless.
8. Rinse in water; stain for five minutes in saffron solution prepared as above; rinse with water.
9. Wash rapidly first with 70% alcohol then with absolute alcohol; clear in xylol and mount.

Results

Nuclei-blue; cytoplasm-varying shades of red; Muscles-pinta elastic fibres-pink, collagen-yellow.

Aniline Blue-Orange G-Acid Fuschsin Method

Solutions required

A. Acid fuchsin 0.5% aqueous

B. Anline Blue-Orange G.

Procedure

Tissues are fixed in Zenker and embedded in paraffin wax, Celloidin of L. V. N.

1. Mount sections on slides and bring down to 90% alcohol ; then with iodine in the usual way to remove mercuric deposits.
2. Bring down to distilled water and stain for one to ten minutes in solution A; then without washing
3. Stain for twenty minutes to one hour or longer in Aniline Blue-Orange G; then remove excess stain with several changes of 95% alcohol.
4. Dehydrate with absolute alcohol; clear in xylol and mount Cristalite.

Note. If Celloidin or L.V.N. sections are used the staining time may be shortened and 95% alcohol should be used for decolourizing and dehydration; terpineol for clearing.

Results

Collagenous fibrils : intense blue. Ground substances of cartilage, bone, mucus, amyloid : varying shades of blue; Nuclei, myoglia, neuroglia fibrils, axis cylinders, fibrin, nucleoli; red. Blood corpuscles and myelin; yellow Elastin fibrils; pale pink or yellow, or unstained; fibriloglia : red or unstained.

Note. By omitting the acid fuchsin the collagenous fibres are more shraply defined.

Mallory Heidenhain Staining Method

Solution required

Phosphotungstic acid crystals A. R.	1 gm
Orange G.	2 gm
Aniline blue water soluble	1 gm
Acid fuchsin	3 gm
Distilled water	200 ml

Procedure

1. Fix pieces of tissue in Zenker-formol for preference, although Bouin's fluid, formalin and alcohol has been used with success.

2. Embed in paraffin was and cut sections 6μ in thickness.
3. Fix sections to slides and remobe wax with xyol.
4. Pass through descending grades of alcohol and if Zenker-formol has been used as the fixative treat with iodine and sodium thiosulphate as usual to remove mercurial precipitate.
5. Take down to tap water.
6. Immerse for five minutes in the staining solution.
7. Wash in running tap water for three to five minutes.
8. Dehydrate rapidly through the usual graded alcohols.
9. Clear in xylol and mount.

Results

Appear to be the same as those listed by Mallory (1938), *i.e.*, collageneous fibrils intense blue. Ground substance of cartilage and bone, mucus amyloid and certain other hyaline substance are stained in varying shades of blue. Nuclei, fibroglia, myoglia and neuroglia fibrils, nucleoli, axis cylinders and fibrin are stained red. Erythrocytes and myelin, yellow. Elastic fibrils are stained pale pink or yellow.

Haematoxylin-Bicbrich Scarlet Picro Aniline Blue Method

Solution required

A. Haematoxylin (Weigert)A.
B. Haematoylin (Weigert) B.

C.	Biebrich scarlet, 0.2% aqueous	10 ml
	Glacial acetic acid	1 ml
D.	Picric acid, saturated aqueous	100 ml
	Aniline blue, water soluble	0.1 gm

E. Acetic acid 1% aqueous.

Procedure

1. Tissues should be fixed in 10% formalin and paraffin sections employed.
2. Stain for five mintues in a freshly prepared mixture consisting equal parts of Weigert's Haematoxylin A and B.
3. Wash in tap water.
4. Stain for three to five minutes in the acetic Poncaeu, S.
5. Rinse in distilled water.
6. Stain for three to five minutes in the picro aniline blue.
7. Wash for three or four minutes in 1% acetic acid solution.

8. Dehydrate in ascending strengths of alcohol and clear in xylol in the usual manner.
9. Mount in acid balsam.

Results

Connective tissue, glomerular basement membrane and reticulum; blue. Muscle and plasma; pink Erythrocytes; bright red.

Mallory's Phosphotungstic Acid Haematoxylin

Solution required

Haematoxylin 10% in absolute alcohol (ripened for three months or longer)	1 ml
Phosphotungstic acid	2 gm
Distilled water	100 ml

Note. If ripened haematoxylin solution is not available, the following artifically ripened stain should be used :

Haematoxylin (dry) 0.1 gm, phosphotungstic acid 2 gm, distilled water 100 ml. potassium permanganate 1% aqueous 1.77 ml.

Procedure

1. Fix in Zenker. Embed in paraffin wax.
2. Bring sections down to distilled water.
3. Treat with iodine to remove mercuric precipitate.
4. Remove iodine with 0.55 aqueous sodium phasphite.
5. Wash thoroughly in running water.
6. Immerse for five to ten minutes in 0.25% potass, per manganate; then wash in tap water.
7. Immerse for ten to twenty minutes in 5% oxalic acid; then wash thoroughly with top water.
8. Stain twelve to twnety-four in haematoxylin solution prepared as above.
9. Wash in tap water; dehydrate with 95% and absolute alcohol.
10. Clear in xylol and mount.

Results

Nuclei, centrioles, achromatic spindles, fibroglia, myoglia, neuroglia, fibrils, fibrin, contractile elements of striated muscle ; blue. Collagen, reticulum ground substances of cartilage and bone : yellowish to brownish red. Coarse elastic fibrils : faint purple.

METHODS FOR RETICULIN

Gordon and Sweets 1939 Method

Solution required

(A) Acid permanganate

Potassium permanganate ($KMnO_4$)	1.0 g
Water	95 ml
3% aqueous H_2SO_4	5 ml

Prepare just before use from a stock 6% aqueous $KMnO_4$. The addition of H_2SO_4 is not necessary.

(B) 1% oxalic acid

Oxalic acid (HOOC. COOH.$2H_2O$)	5 g
Water	to 500 ml

Keeps indefinitely

(C) Iron alum

Iron alum ($NH_4Fe(SO_4)_2.12H_2O$)	10 g
Water	to 500 ml

Prepare on the day it is to be used. Alternatively, use 4% aqueous ferric chloride ($FeCL_3.6H_2O$), which keeps indefinitely.

(D) Ammoniacal silver solution

Stock solutions

1. 10% aqueous silver nitrate.
2. Ammonium hydroxide (28% NH_2).
3. 4% aqueous sodium hydroxide.

Working solution

Add ammonium hydroxide drop by drop to 10 ml of 10% $AgNO_2$ until the brown precipitate of Ag_2O is almost (not quite) resolved. Add 7.5 ml of 4% NaOH, followed by a few more drops of ammonium hydroxide, until the realy formed precipitate is just dissolved. Be careful not to add too much ammonia. Swirl the solution by a few seconds after adding each drop, since disolution of the precipitate is not quite instantaneous. Make up to 100 ml with water. *This solution should be made just before use and discarded afterwards by washing it down the sink with plenty of water.* Ammoniacal silver solutions decompose on evaporation to form explosive "fulminating silver", a mixture of silver amide and silver nitride.

(E) Reducer

Neutralized formalin (40% HCHO which has stood over marble):	10 ml	Prepare just
Water	90 ml	before using

(F) Yellow gold chloride.

Sodium tetrachloroaurate ($NaAuCl_4.2H_2O$)	1 g	Keep for
Water	500 ml	several months

This solution may be re-used repeatedly

(G) Sodium thiosulphate

Sodium thiosulphate ($Na_2S_2O_3.5H_2O$) :	25 g	Keeps for
Water:	to 500 ml	several months

Procedure

1. De-wax and hydrate paraffin sections.
2. Oxidize for 1 min in acid permanganate (solution A).
3. Wash in water.
4. Immerse in 1% oxalic acid (solution B) until the sections are white. Usually about 30s.
5. Wash in water (three changes).
6. Treat with iron alum (solution C). 10 min.
7. Wash in water (three changes).
8. Immerse slides in the ammoniacal silver solution (solution D) 5-10s.
9. Rinse in water (once only.)
10. Place in formaldehyde reducer (soultion E), 30s.
11. Wash in water (three changes).
12. Tone in 0.2% yellow gold chloride (solution F), 2 m.
13. Wash in water (two changes).
14. Immerse in sodium thiosulphate (solution G), 3 min,
15. Wash in water (three chages).
16. Dehydrate through graded alcohols, clear in xylene and cover.

Result

Reticulin—black.Other elements in shades of grey.

Methods for Elastin

Weigert Elastin Staining Method

Preparation of the staining solution

Triturate gm of Weigert elastin stain and 5 gm clean, dry silver sand with 100 ml absolute alcohol and 2 ml pure hydrochloric acid until all the stain has gone into solution; then filter.

Note: The staining solutions deteriorates after two or three weeks.

The nuclei may be stained with Orth's lithium carmine prior to the following procedure if no other counterstain is desired.

Procedure

1. Sections are brought down to 90% alcohol and stained one half hours according to depeth of staining desired. The slides should be stained in a jar or in a Petri dish, sections face downwards to prevent a deposit forming on the sections.
2. Wash off excess stain with 95% alcohol, and if necessary differentiate in acid alcochol for a few minutes.
3. Wash quickly with 70% alcohol: then thoroughly with water.
4. Counterstain with Van Gieson, Ehrlich haematoxylin or Safrainin for about five minutes.
5. Differentiate, if necessary, in 95% alcohol.
6. Dehydrate; clear in xylol an dmount.

Note. If Celloidin or L.V.N. sections are used clear in origanum oil or in terpineol after 95% alcohol.

Results

Elastic fibres, dark blue or black, Nuclei, brilliant red (if Orth's carmine is used) or bluish black (with haematoxylin). Collagen, pink to red : other tissue elements, yellow (if Van Gieson is used).

Elastin-Trichrome Staining Method

Solutions required

A. *Weigert's elastin stain*

Weigert's elastin stain powder	1 gm
Hydrochloric acid, cone, pure	2 ml
Absolute alcohol	100 ml

Dissolve the stain by boiling for two minutes in flask, plugged lightly with cotton-wool, on a water bath. Allow to cool; then filter; make the volume upto 100 ml with absolute alcohol; then add the acid. Alternatively, the solution may be prepared as described.

Note. This solution deteriorates after three or four weeks.

B. Ehrich haematoxylin.

C. *Poneau-acid fuchsin(Masson):*

Acid fuchsin	0.3 gm
Ponceau de xylidine	0.7 gm
Distilled water	100 ml
Glacial acetic acid	1 ml

D. Phosphotungstic acid 3% aqueous

E. Light Green 1% aqueous.

Procedure

1. Paraffin section are mounted on slides and brought down to distilled water in the usual manner; then immersed in Weigert's elastin stain in a staining jar for one hour.
2. Wash rapidly in acid alcohol; then dehydrate and differentiate in absolute alcohol until the sections appear only faintly red.
3. Immerse in 70% alcohol, followed by distilled water.
4. Stain in Ehrlich haematoxylin for eight to ten minutes; then differentiate in water for five minutes.
5. Stain in Ponceau-acid fuchsin for five minutes.
6. Wash thoroughly in 3% phosphotungstic acid; then immerse in the phosphotungstic acid for ten minutes.
7. Wash thoroughly in distilled water; then stian with Light Green for two to five minutes; then without washing.
8. Flood the preparation with 1% acetic acid and allow it to act for three minutes; pour off excess ; then without washing.
9. Dehydrates; clear; mount in DP.X.

Results

Elastic tissue stained blue-black; smooth muscle, red; collagen, green.

Congo Red-Aniline Blue-Orange G Method

Solutions required

A. Aluminium chloride, 2% aqueous

B. Congo Red	2 gm
Sodium citrate	2.5 gm
Glycerin	1 ml

Distilled water	94 ml
C. Aniline Blue, aqueous	1.5 gm
Orange G	2.25 gm
Resorcinol	3 gm
Phosphomolybdic acid 1% aqueous	100 ml

Procedure

Tissues should be fixed in 10% formalin, and frozen sections should be empolyed.

1. Wash sections in water; then immerse them in solution A for ten minutes.
2. Wash with water and drain; then stain in the Congo Red solution for ten minutes.
3. Wash with tap water; then plunge the slide into a dish of tap water and agitate it there for ten seconds.
4. Wash again with tap water; then stain from five to ten minutes in the Aniline Blue-orange G solution (solution C above).
5. Rinse carefully in tap water; drain well and blot.
6. Dehydrates in absolute alcohol; clear in origanum oil; wash in xylol and mount.

Results

Elastic fibres : bright red. Fibrin : dark blue. Erythrocytes: yellowish orange.

Orcinol-New Fuchsin Method

Solutions required

A. New fuchsin	2 gm
Orcinol	4 gm
Distilled water	200 ml
Boil for five minutes; then add:	
Ferric chloride ($FeCl_2$), 15%	25 ml
and boil for a further period of five minutes.	

Allow the solutions to stand until cold; then collect the precitpitate by filtration : washing and drying is stated to be unnecessary.

Dissolve the precipitated orcinol-new fuchsin in 100 ml of 95% alcohol and use this as the elastin stain.

Procedure

1. Material may be fixed in 10% formalin or saturated aqueous mercuric chloride or Zenker, etc.

2. Fix sections to slides, dewax and pass through absolute alcohol as usual.
3. If a mercury containing fixative has been used, treat sections for the removel of mercurial precipitate by the standard method, afterwards rinsing with 90% alcohol.
4. Stain in the orcinol new fuchsin for fifteen minutes at 37°C
5. Differentiate for five minutes in each of three changes of 70% alcool.
6. Dehydrate in absolute alcohol ; clear in xylol, and mount in DPX,or Cristalite, or Clear mount, etc.

Results

Elastic fibres : deep violet. Collagen : unstained.

Notes. The authors reported that they used the stain on human skin and aorta, and the following tissues of the mouse : liver, kidney spleen, stomach, duodenum colon, pancreas, heart, testes, seminal vesicles, ovary, uterus, aorta, pituitary, salivary, gland bone striated muscle and thyroid, and in all cases only elastic tissue was stained.

Orcein-Aniline Blue-Orange G.

Solution required

A.	Orcein	1 gm
	Alcohol 70%	100 ml
	Hydrochloric acid, conc.	0.6 ml
B.	Alcohol 50%	49 ml
	Hydrochloric acid, cone.	0.5 ml
C.	*Mallory's Aniline Blue-Orange G.*	0.5 gm
	Orange G	2 gm
	Phosphomolubdic acid 1%	100 ml

Procedure

1. Fix material in Bouin and embed in paraffin wax.
2. Sections, about 8μ in thickness are fixed to slides, dewaxed with xylol and passed through the usual descending grades of alcohol to distilled water.
3. Stain for one and a half hours in the orcein solution in a closed staining jar.
4. Differentiate with solution B, controlling under the microscope, until most of the pink is extracted from the sections.

   Note. The duration of the differentiation will vary according to the nature of the material and to the thickness of the sections.

5. Wash thoroughly with running tap water.
6. Wash with distilled water.
7. Immerse in solution C diluted with an equal volume of distilled water, for one to three minutes.
8. Rinse with 95% alcohol.
9. Rinse with two lots of solution D.
10. Rinse quickly with absolute alcohol.
11. Clear in xylol and mount.

Results

Elastic fibres: red, Collagen : blue. Muscle fibres : pale orange to dirty yellow. Keratinized material : bright yellow.

Some General Staining Methods for Connective Tissues

Haematoxylin-Picro Fuchsin Method

Solution required

A.	Distilled water	47.5 ml
	Ferric chloride, hydrated 4% aqueous	2 ml
	Haematoxylin 10% in absolute alcohol	0.4 ml
B.	Picric acid, saturated, aqueous	
	Acid fuchsin 1%aqueous	0.5 ml
C.	Picric acid, saturated in absolute alcohol.	

Procedure

1. Tissues are fixed in Bouin and embedded in paraffin wax.
2. Sections about 8μ in thickness are fixed to slides, dewashed with xylol and taken through the usual decending graders of alochol to distilled water.
3. Stains for two to three minutes in solution A.
4. Differentiate and counterstain for about ten to fifteen seconds in solution B, controlling under the microscope, until only the nuclei are stained a greyish colour with the haematoxylin.
5. Rinse immediately in distilled water.
6. Dehydrate by dripping solution C onto the slide.
7. Clear with Terpineol.
8. Mount direct with Michrome mountant, or rinse with xylol, then mount with Clearmount or Cristalite.

Results

Chromatin : black to grey. Muscle : yellow. Connective tissue : red. Keratinized regions : bright yellow. Cytoplasm: yellow.

Methyl Violet-Pyronin-Orange G Method

Solution required

A.	Methyl violet 6B (Jensen) 1% aqueous	25 ml
	Pyronin B 10% aqueous	10 ml
	Distilled water	65 ml
B.	Acetone	100 ml
	Orange G aqueous 2%	about 10 ml

Add orange G solution drop by drop to the acetone, with shaking until the flocculent precipitate formed just redissolves with further addition of orange G solution.

Procedure

1. Small pieces of tissue are fixed in acetic-alcohol or in mercuric choride.
2. Wash; dehydrate; clear, embed in paraffin wax.
3. If mercuric chloride has been used for fixation treat sections for the removal of mercuric precipitate by the standard method.
4. Take sections down to distilled water.
5. Immerse for two minutes in the methyl violet pyronin (Solution A above).
6. Pour off excess stain and carefully wipe the slide dry.
7. Flood the preparation with acetone-orange G solution.
8. Pour off after a few seconds.
9. Flood the preparation with a fresh lot of acetone-orange G solution and pour off after a few seconds.
10. Wash quickly in pure acetone.
11. Rinse with two lots of xylol.
12. Mount in balsam.

Results

Cytoplasm, red; chromatin, violet : keratin, violet; connective tissue, yellow.

Nephthol Green-B-Haematoxylin Method

Solutions required

A. Weigert's haematoxylin, A.

B. Weigert's haematoxylin. B,

C. Eosin, yellowish, 1% in tap water.

D. Ferric chloride, hydrated 10%

E. Naphthol Green B 15 aqueous

F. Equal volumes of acetone and xylol.

Procedure

1. Paraffin sections are mounted on the slide and broughts down to distilled water in the usual manner.
2. Stain for six minutes in a freshly prepared mixture consisting of equal volumes of Weigerts haematoxylin A and B,
3. Wash throughly in tap water; then stain for three minutes in the eosin solution.
4. Wash in tap water; then immerse in the ferric chloride solution for five minutes
5. Rinse well in distilled water; then stain for five minutes in the naphthol green solution.
6. Differentiate for two or three minutes in 1% acetic acid.
7. Drain well; then dehydrate with acetone, afterwards clearing in acetone-xylol (as above); then mount.

Results

Connective tissues, green; muscle and cytoplasm, pink

Aniline-Crystal Vinlet-Lithiun Carmine-Iodine Method

Solutions required

A. Lithium carmine

B. Crystal violet 1 gm

Aniline oil 3 ml

Absolute alcohol 10 ml

Dissolve and filter.

C. Crystal violet 2% aqueous

D. Solution B 3 ml

Solution C 27 ml

This mixture should be prepared immediately before use.

E. Gram's iodine.

F. Aniline oil 1 volume

Xylol 1 volume

Procedure

1. Fix material in absolute alcohol, Carnoy or alcoholformalin and embed in paraffin wax.
2. Fix sections to slides; de-wax and pass through descending grades of alcohol down to distilled water in the usual way.

3. Stain in the lithium carmine solution for two to five minutes.
4. Wash thoroughly in distilled water.
5. Immerse in Solution D for five to ten minutes.
6. Rinse in distilled water; drain well and blot carefully
7. Cover with Gram's iodine solution and allow the stain to act for five to ten minutes.
8. Pour off the excess iodine solution and blot carefully with filter paper.
9. Differentiate with the aniline xylol solution until no more purple colouration comes out.
10. Drain, and blot carefully
11. Rinse with several changes of xylol.
12. Mount in balsam or in cristalite.

Results

Fibrin and Gram-positive organisms are blue to blue-black while nuclei are red.

Alizarin Red-S method (Dawson's Method)

Solutions required

A. Potassium hydroxide 1% aqueous

B. Alizarin Red, S 0.1 gm
   Potassium hydroxide 10 gm
   Distilled water 1 liter

C. *Mall's solution :*
   Glycerin 20 ml
   Distilled water 79 ml
   Potass, hydroxide 1 gm

Procedure

1. Whole specimens are fixed in 95% alcohol for at least three days.
2. Transfer to acetone and leave for several days to dissovle out the fats which would otherwise stain intensely and obscure the view of the bony structures.
3. Wash well with 95% alcohol; then immerse in 95% alcohol for twenty-four hours.
4. Immerse in Solution A from one to seven days, according to the size of the specimen, until the bones are clearly visible through the muscle.

5. Transfer to solution B until the bones are stained the desired depth of colour; this takes from one to seven days, and the solution should be changed on the fourth day.
6. Clear in Solution C until no more colour comes out.
7. Pass into a mixture of equal parts of glycerin and water, and continue through increasing strength of glycerin.
8. Store in pure glycerin.

Results

Bones are stained red; soft tissue, transparent and unstained.

Benda's Method (for Nervous Tissues)

Solutions required

A. Nitric acid, conc.

Distilled water — 1 volume

B. Potassium dichromate 2% — 10 volume

C. Chromic acid 1%

D. Iron alum 4%

E. Alizarin Red, S, saturated in

Absolute alcohol — 1 ml

Distilled water — 90 ml

F. Toluidine Bue 0.1% aqueous

Procedure

1. Material is fixed in 90-95% alcohol for at least two days.
2. Pieces, which must not be thicker than 0.5 cm, are immersed in Solution A for twenty-four hours.
3. Transfer to solution B for twenty-four hours
4. Transfer to solution C for forty-eight hours; then wash in water for twenty-four hours.
5. Dehydrate in teh usual manner.
6. Clear in beechwood creosote for twently four hours; then in benzol for twenty-four hours.
7. Embed in paraffin was via four graded mixtures of paraffin wax and benzol, the first at room temperature, the second at 38^{0}C, the third at 42^{0}C and the fourth at 45^{0}C.
8. Mount sections on slides; bring down to distilled water, mordant sections on slides with Solution D for twenty-four hours then wash thoroughly in water.

9. Stain for two hours with Solution E; then rinse in tap water.
10. Flood slides with Solution F and warm gently until vapour is given off ; or stain at room temperature for 24 hours.
11. Rinse in 1% acetic acid; then dry by blotting carefully.
12. Pass through absolute alcohol; then differentiate for about ten minutes in beechwood creosote; dry by bloting carefully, wash with xylol, and mount

Result

Soft tissue-Transparent ; Osseous tissue-deep blue.

Ammoniacal Silver Carbonate Method (for tumor cells)

Solutions required

A. Pyridin, pure — 2 volumes
Glycerin, pure — 1 volumes

B. *Ammoniacal Silver Carbonate*

Ammonia solutions is added drop by drop to 10 ml silver nitrate 10.2% until the precipitate formed is almost redissolved, leaving a slightly opalescent solution to which is then added 10 ml sodium carbonate 3.1% soultion and sufficient distilled water to make the volume up to 100 ml.

C. *Reducing solution*

Sodium carbonate anhydrous	1 gm
Formalin	1 ml
Distilled water	103 ml

D. Brown gold chloride 0.2% aqueous

E. *Intensifying solution*

Oxalic acid 2% aqueous	100 ml
Formalin	1 ml

F. Sodium hyposulphite 10% aqueous.

Procedure

The material is fixed in 10% formalin or in Bouin and embeded in paraffin wax.

1. Bring sections down to distilled water and immerse in Solution A for twenty-four hours.
2. Wash with 95% alcohol, then with distilled water.
3. Immerse in Solution B for two and a half hours at 40°C.
4. Wash with distilled water; then reduce in Solution C for five minutes, afterwards washing in tap water.

5. Tone for five minutes in Solution D at 30°C; then wash in tap water.
6. Intensity by immersing in Solution E for five minute; then rinse in tap water.

   N. B. The above stages must be carried out in the dark-room.
7. Fix in Solution F.

   (Note. Fixation should be completed in fifteen to twenty minutes).
8. Wash in tap water; dehydrate; clear and mount.

Results

Tumour cells; reddish to greyish violet. Vascular reticulum; black.

Bismarck Brown-Methyl Green Method (for mucin, cartilage)

Solution required

A. Bismarck brown 1% aqueous

B. Methyl green 0.5% aqueous.

Procedure

Tissues are fixed or Zenker and embedded in paraffin wax.

1. Sections are brought down to distilled water; then stained five to ten minutes in solution A.
2. Wash with 95% alcohol.
3. Stain with Solution B until the preparation appears dark green to the naked eye.
4. Dehydrate with 95% and absolute alcohol; then clear in xylol, and mount.

Results

Cartilage : dark brown, Mucin : light brown. Nuclei of all cells : green.

Notes. The authors state that the method is essentially the same as List (1885).

Cresylfast Violet-Toluidine Blue-Thionin Method (Ehrlich Method) (for nerve cells)

Solution required

A.	Cresylfast violet, CNS	2 gm
	Toluidine blue	1 gm
	Thionin (Ehrlich)	0.5 gm
	Ethyl Alcohol 30%	200 ml

B. Distilled water 200 ml

Sulphuric or nitric acid, conc 0.5 ml

Procedure

1. Formalin fixed material is embedded in paraffin wax and section 4μ in thickness are fixed to slides with glycerin albumen.
2. Remove wax with xylol.
3. Rinse with absolute alcohol
4. Pass through 95% alcohol.
5. Pass through 80% alcohol.
6. Immerse slides for five toten seconds in the staining solution at 80-90°C
7. Differentiate for one second in solution B.
8. Dip and agitate slides in a beaker of cold distilled water for one second.
9. Differentiate further in 80% and 95% alcohol for one to two seconds in each.
10. Immerse in 80% alcohol for one second.
11. Dip and agitate the slides in the still warm solution A for one to two seconds.
12. Return to 80% alcohol for one second.
13. Repeat steps 11 and 12.
14. Rinse in distilled water.
15. Dehydrate by immersing for one second in each of 80%, 95% and absolute alcohol.
16. Immerse in xylol for one minute.
17. Immerse in a fresh lot of xylol for three minutes.
18. Mount and examine.

Reaults

Neurons stand out distinctly against a pale background, and can be followed for a considerable distance. The cytons are stained dark purple emphasizing the blue tint, while the dendrite and axon processes and endings present somewhat lighter shade, bluish to reddish. Granules in the cell-body as well as inthe protoplasm processes appear purple or reddish. Nuclei and nucleoli are well differentiated.

18

THE MICROSCOPES

The microscope is the most commonly used piece of apparatus in the laboratory and yet it is probably the instrument about which least is known by its users. It is generally thought that the microscope can be used effectively without any knowledge of its limitations or construction, but this is, of course a complete misconception. An ill-adjusted badly illuminated microscope can, when one is using high-power objectives, give completely misleading information as to the structure of an object. For this reason it is advisable to gain a knowledge of how the magnified images are produced by the microscope before attempting to assess the information obtained by its use. Bacteria and viruses are so small that they cannot be seen with the naked eye. They must be greatly magnified before they can be clearly seen and studied. The use of a microscope is, therefore, absolutely indispensable to the bacteriologist and to the biologist in general.

A microscope may be defined as an optical instrument, consisting of a lens or a combination of lenses, for making enlarged or magnified images of minute objects. A simple microscope, or a single microscope, consists merely of a single lens of magnifying glass held in a frame, usually adjustable, and often provided with a stand for conveniently holding the object to be viewed and a mirror for reflecting the light. A compound microscope differs from a simple one in that it consists of two sets of lenses, one known as an objective and the other as an eyepiece, commonly mounted in a holder known as a body tube. The one nearest the specimen, called the objective, magnifies the specimen a definite amount. The second lens system, the eyepiece, further magnifies the image formed by the objective, so that the image seen

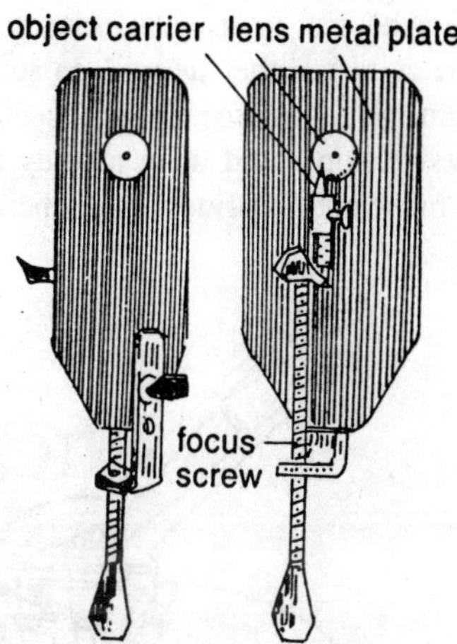

Fig. 18.1. Leeuwenhoek's microscope.

by the eye has a magnification equal to the product of the magnifications of the two systems. The individual or initial magnification of the objectives and eyepieces is engraved on each such part. Accurate focusing is attained by a special screw appliance known as a fine adjustment. Compound microscope give much greater magnifications than simple mictoscopes and are necessary for viewing and examining such minute objects as bacteria. Every user of the microscope should first understand the principles involved in order that the instrument may be employed to the greatest advantage. As Sir A. E. Wright

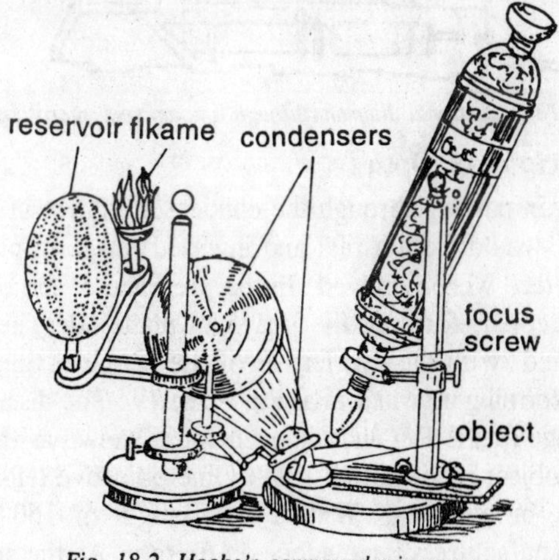

Fig. 18.2: Hooke's compound microscope.

stated: "Every one who has to use the microscope must decide for himself the question as to whether he will do so in accordance with a system of rule of thumb, or whether he will seek to supersede this by a system of reasoned action based upon a study of his instrument and a consideration of the scientific principle of microscopical technique."

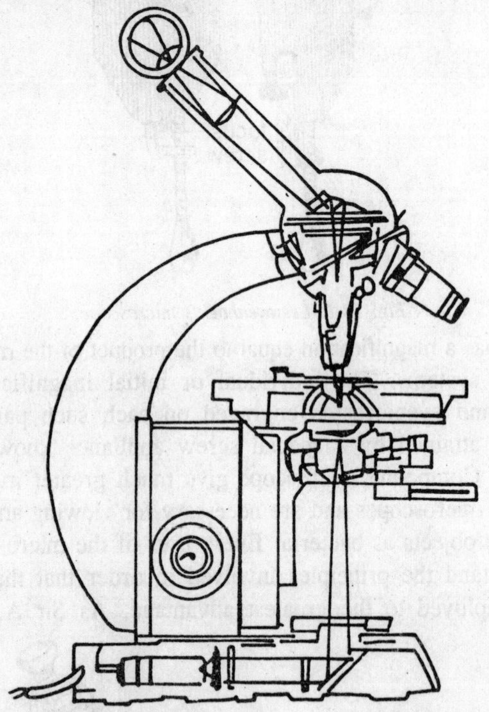

Fig. 18.3. Sectional diagram through a compound microscope.

General Principles of Optics

The light, in passing through the condenser, object in plane I, and objective lens, would from a real and inverted image in plane II if the ocular or eyepiece were removed. In the presence of the ocular F, the rays are intercepted, forming the image in plane III. The real image is then examined, with the eye lens E of the ocular acting as a single magnifier and forming a virtual image in Plane IV. The distance between the virtual image (plane IV) and the eyepoint is known as the projection distance. The object is magnified first by the objective lens and second by the ocular, or eyepiece. With a tube length of 160 mm. (Most microscope manufacturers have adopted 160 mm. As the standard tube

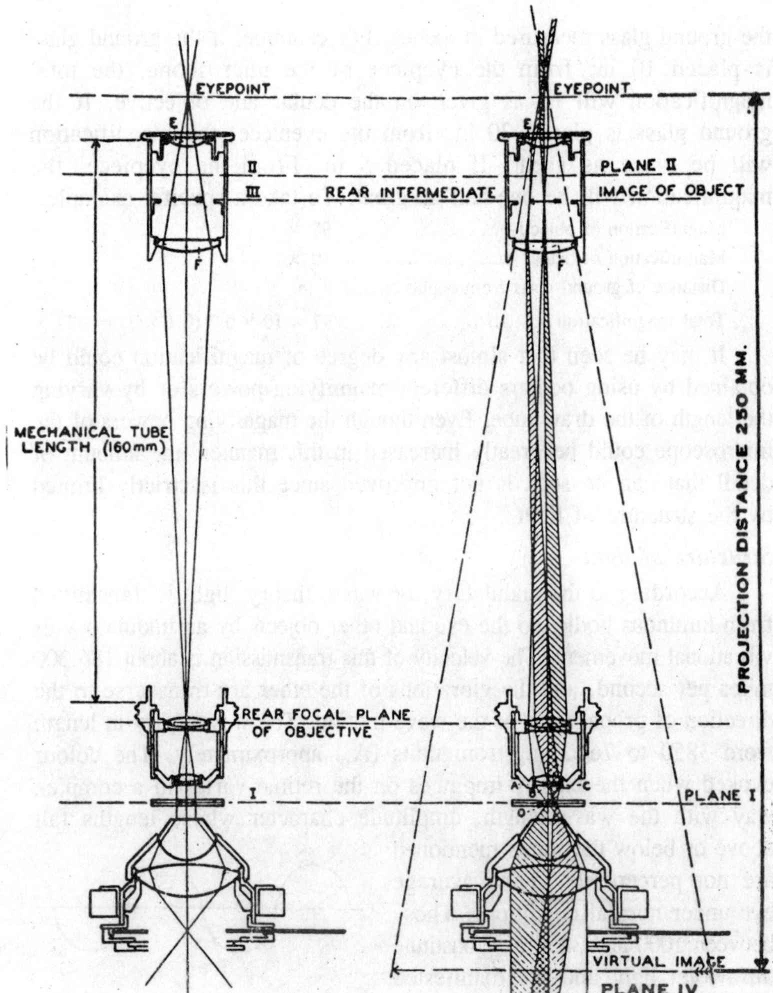

Fig. 18.4. Path of light through a microscope.

length), the total magnification of the microscope is equal to the magnifying power of the objective lens multiplied by the magnifying power of the ocular.

The above magnifications are obtained on a ground glass placed 10 in. from the ocular of the microscope. After the microscope has been set at the proper tube length, the total magnification may be computed by multiplying the magnifying power of the objective by that of the eyepiece and by one-tenth of the distance from the eyepiece to the ground glass measured in inches. For example, if the ground glass is placed 10 in. from the eyepiece of the microscope, the total magnification will be as given on the ocular and objective. If the ground glass is placed 20 in. from the eyepiece, the magnification will be twice as great. If placed 5 in. From the eyepiece, the magnification will be one-half as great. To take a specific example:

Magnification of objective .... 97 ×
Magnification of ocular .... 10 ×
Distance of ground glass from ocular .... 7 in.
Total magnification................................ 97 × 10 × 0.7 (0.1 × 7) = 679 ×

It may be seen that almost any degree of magnification could be obtained by using oculars different magnifying powers or by varying the length of the draw tube. Even though the magnifying powers of the microscope could be greatly increased in this manner, the amount of detail that can be seen is not improved since this is strictly limited by the structure of light.

Structure of light

According to the undulatory, or wave, theory, light is transmitted from luminous bodies to the eye and other objects by an undulatory or vibrational movement. The velocity of this transmission is about 186,300 miles per second, and the vibrations of the ether are transverse to the direction of propagation of the wave motion. The waves vary in length from 3850 to 7600 angstrom units (A.) approximately. The colour evoked when the energy impinges on the retime varies in a complex way with the wave length, amplitude character whose lengths fall above or below the limits mentioned are not perceptible to the average eye under normal conditions. Those between 1000 and 3850 A. Constitute ultraviolet light and are manifested by their photographic or other chemical action. Those exceeding 7600 A. are the infrared waves and are detected by their thermal effects. When a beam of white light is passed through a prism, a spectrum is obtained in which several colours form a series from deep red through orange, yellow, green, blue, and indigo to deepest violet.

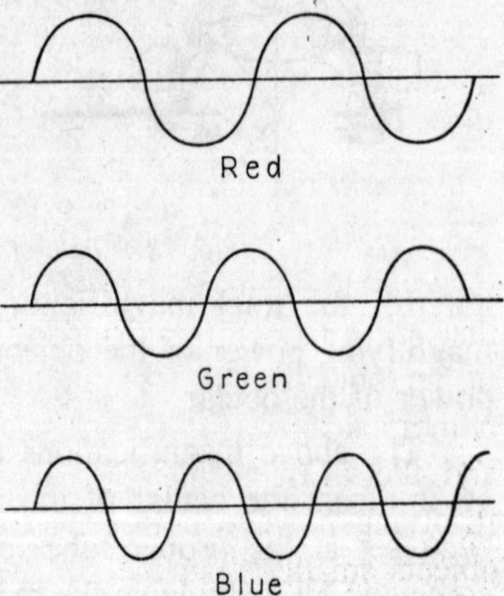

Fig. 18.5. Wave length of light of different colours.

The wave lengths of the various colours are different; red shows the longest and violet the shortest waves of the visible spectrum. The length of a light wave is the distance from the crest of one wave to the crest of the next. The unit of measurement is the angstrom unit (A.), which is equal to 1/10,000,000 mm., or to approximately 1/250,000,000, in. The visible spectrum, together with the corresponding wave lengths of the light rays in angstrom units, may be represented. Visible light waves, ranging in length from 4000 to 7000 A., may be roughly divided into three portions; blue-violet, from 4000 to 5000 A.; green, from 5000 to 6000 A.; red, from 6000 to 7000 A.

Objectives

The objective is the most important lens on a microscope because its properties may make or mar the final image. An objective capable of utilizing a large angular cone of light coming from the specimen will have better resolving power than an objective limited to a smaller cone of light. The chief functions of the objective lens are (1) to gather light rays coming from any point of the object, (2) to unite the light in a point of the image, and (3) to magnify the image. There are three major types of objectives, namely, achromatic, fluorite, and apochromatic.

The achromats are the simplest in construction and the least expensive. They are adequate for most purposes, Correction for both colour and spherical aberration is quite good in the lower-power objectives, but the control of aberrations becomes more difficult as the power is increased. Aberrations are largely eliminated by the use of fluorite (semiapochromatic) objectives and, especially, the apochromats. The latter are more highly corrected with respect to aberrations than any other type of objective and are preferred for the most critical work.

Numerical aperture

The resolving power of an objective may be defined as its ability to separate distinctly two small elements in the structure of an object that are a short distance apart. The measure for the resolving power of an objective is the numerical aperture (N.A.). The larger the numerical aperture, the greater the resolving power of the objective and the finer the detail it can reveal. Since the limit of detail or resolving power of an objective is fixed by the structure of light, objects smaller than the smallest wave length of visible light cannot be seen. In order to see such minute objects, it would be necessary to use rays of shorter wave length.

Invisible rays, such as ultraviolet light, are shorter than visible rays but since they cannot be used for visual observation, their usefulness is limited. The image of an object formed by the passage of light through a microscope will not be a point but, in consequence of the diffraction of the light at the diaphragm, will take the form of a bright disk surrounded by concentric dark and light rings. The brightness of the central disk will be greatest in the center, diminishing rapidly toward the edge.

The image cone of light composed of a bright disk surrounded by concentric dark and light rings is spoken of as the antipoint. If two independent points in the object are equidistant from the microscope lens, each will produce a disk image with its surrounding series of concentric, dark and light rings. The disks will be clearly visible if completely separated, but if the images overlap they will merge into a single bright area, the central portion of which appears quite uniform. The two disks will not, therefore, be seen as separate images. It is not known how close the centers of the images can be and still be seen as separate antipoints.

The minimum distance between the images of two distinct object points decreases as the angle of light *AOC*, coming from the object, *O*, increase. The angle formed by the extreme rays is known as the aperture of the objective. The ability of the objective lens system to form distinct images of two separate object points is proportional to the trigonometric sine of the angle. The latter, then, is a measure of the resolving power of the objective. Actually, however, the sine of angle *AOB* is used, which is just one-half of angle *AOC*. This is usually referred to as sin μ. Since the sine of an angle may be defined as the ratio of the side opposite the angle in a right-angled triangle to the hypotenuse, then

$$\sin\mu = \frac{AB}{AO}$$

The light in passing through the objective is influenced by the refractive index *n* of the space directly in front of the lens. This is another factor that affects the resolving power of an objective. The two factors, refractive index *n* and sin μ, may be combined into a single expression, the numerical aperture, which may be expressed as follows:

$$\text{N.A.} = n \sin \mu$$

Importance of N.A.

If a very narrow pencil of light is used for illumination, the finest detail that can be revealed by a microscope with sufficient magnification is equal to

$$\frac{\text{w.l.}}{\text{N.A.}}$$

where w. l. is the wave length of the light used for illumination and N. A. is the numerical aperture of the objective. The resolving power of the objective is proportional to the width of the pencil of light used for illumination. This means that the wider the pencil of light, the greater the resolving power. The maximum is reached when the whole aperture of the objective is filled with light. In this instance, the resolving power is twice as great. The finest detail that the objective can reveal is now equal to

$$\frac{\text{w.l.}}{2\text{ N.A.}}$$

For example, the brightest part of the spectrum shows a wave length of 5300 A. An objective having a numerical aperture equal to 1.00 will resolve two lines separated by a distance of 5300 A./1.00 = 5300 A. (48,000 lines to the inch) if a very narrow pencil of light is used, and 5300 A./(2 × 1.00) = 2650 A. 95,000 lines to the inch) if the whole aperture of the objective is filled with light. From the above, it is evident that the maximum efficiency of an objective is not reached unless the back lens is filled with light. This may be ascertained by removing the eyepiece from the microscope and viewing the back lens of the objective with the naked eye. If the back lens is completely filled with light, the efficiency will then be according to the numbers engraved on the objective.

Resolving power

The shorter the wave length of light, the finer the detail revealed by the objective. With an objective having an N. A. Of 1.00 and a yellow filter (light transmission of 5790 to 5770 A.), it is possible to see about 88,000 lines to an inch; with a green filter (light transmission of 5460 A.), about 95,000 lines to an inch; with a violet filter (light transmission of 4360 A.), about 115,000 lines to an inch; and with ultraviolet light (light transmission of 3650 A.), about 140,000 lines to an inch.

Immersion objectives

When a dry objective is used, an air space is present on both sides of the microscope slide and cover slip. The largest cone of light coming from *O* that could possibly be used is 180° in air, which is equal to an angle of about 82° in the glass. This corresponds to a numerical aperture 1.0. In actual practice, however, these figures become

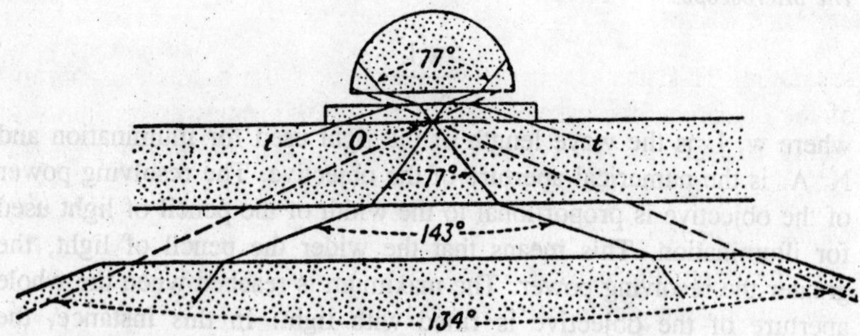

Fig. 18.6. Passage of light through an object on a glass slide using dry and immersion objectives.

143 and 77° respectively, owing to the fact that the air space must be wide enough to correspond to a practical working distance of the objective. Rays of greater angular aperture than 82° in glass, which originate at the object point *O* by diffraction, will be completely reflected at the upper surface of the cover slip *t*. The refractive index *n* of the air is equal to 1.0.

Table 18.1

Medium	*Refractive index at 25°C*
Water	1.33
Mineral (Paraffin) oil	1.47
Cedarwood oil	1.51
Sandalwood oil	1.51
Shillaber's immersion oil	1.52
Balsam	1.53
Crown oil	1.55

If the air space between the cover slip and the objective is filled with a fluid having a higher refractive index, such as water (n = 1.33), or, what is still better, a liquid having a refractive index approaching that of glass, such as cedarwood oil (n = 1.51), angles greater than 82° are obtained. Numerical apertures greater than 1.0 are realized by this method. Cedarwood oil causes the light ray to pass right through the homogeneous medium, with the result that a cone of light of about 134° is obtained, which corresponds to a numerical aperture of 1.4. Finer detail can, therefore, be resolved by this procedure. With an oil-immersion objective and a numerical aperture of 1.4, two lines as close to ether as 1/100,000 in. (0.2 μ) can be separated. This means, then, that the greater the numerical aperture of the objective, the greater will be its resolving power or ability to record fine detail. The refractive indexes of a number of media that have been employed for immersion objectives are given in Table 18.1

Depth of focus

The depth of focus is known also as the depth of sharpness or penetration. The depth of focus of an objective depends upon the N A. and the magnification, and is inversely proportional to both. This means that the higher the N.A. and the magnification, the lower the depth of focus. Therefore, high-power objectives must be more carefully focused than low-power objectives. These conditions cannot be changed by the optician.

Equivalent focus

Objectives are sometimes designated by their equivalent focal lengths measured in either inches or millimeters. An objective designated by an equivalent focus of $^1/_{12}$ in., or 2 mm., means that the llens system prodces a real image of the object of the same size as is produced by a simple biconvex or converging lens having a focal distance of $^1/_{12}$ in., or 2 mm.

Working distances of uncovered objects

If the object on a glass slide is not covered with a cover lip, the working distance may be defined as the distance between the front lens of the objective and the object on the slide when in sharp focus. The working distance is always less than the equivalent focus of the objective. The working distance may be determined easily by noting the number of complete turns of the micrometer screw (fine adjustment) required to raise the objective from the surface of the slide, where the object is located, to a point where the microscope is in sharp focus.

To take a specific example :

Each turn of the micrometer screw = 0.1 mm,

Number of turns required to bring object in sharp focus = 6

Working distance = 6 × 0.1 = 0.6 mm.

Working distance of covered objects

If the object is covered with a cover slip, the free distance from the upper surface of the cover slip to the front of the objective will be less than in the case of an uncovered object. It is obvious from this that if the cover glass is thicker than the working distance of the objective, it will be impossible to get the object in focus. On the

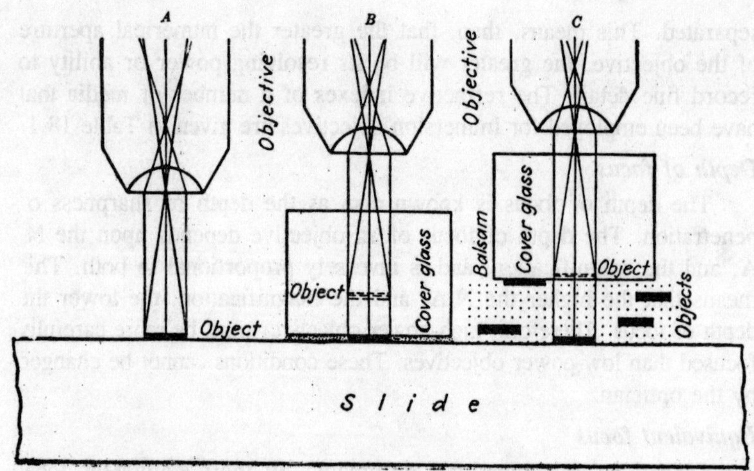

Fig. 18.7. Working distance of an objective. A, object not covered with a cover slip. B, C, object covered with a cover slip.

other hand, if the glass is thin it will be possible to get the object in focus, but the focus of the microscope on a covered object will be different from that on an uncovered object. It follows from this that an object covered with a glass cover slip or other highly refractive body will appear as if raised, and the amount of elevation will depend upon the refractive index of the glass or other medium covering the object. Also the greater the refraction of the covering body, the more will be the apparent elevation. The apparent depth of the object below the surface of the covering medium may be calculated by taking the reciprocal of its index of refraction. For example, if a glass cover slip is used, it will have an index of refraction of 1.52. The reciprocal of this figure is 1/1.52 = 2/3, approximately. This means that the apparent depth of the object is only two-thirds its actual depth. The working distance of covered objects may be determined by noting the number of complete turns of the micrometer screw fine adjustment) required to raise the objective from the surface of the cover slip to a point where the objective is in sharp focus. To take a specific example:

Each turn of micrometer screw = 0.1 mm.

Number of turns required to bring object in sharp focus = 3.5

Working distance = 3.5 × 0.1 = 0.35 mm.

Aberrations in objectives

Perfect lens systems have not yet been designed. All lens systems have aberrations to a greater or lesser degree, depending upon the

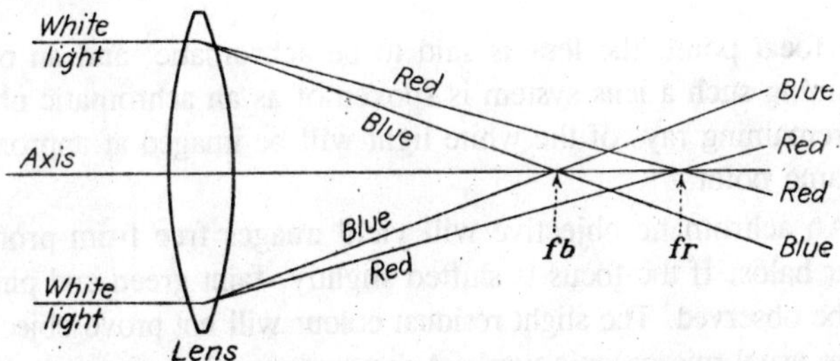

Fig. 18.8. Chromatic aberration with white light. White light, in passing through a lens, is dispersed into its constitutent colours. The red or long waves are refracted less than the blue or short waves. The blue rays (fb) cross the optical axis of the lens before the red rays (fr). The blue light will focus nearer the lens than the red light.

skill of the designer and the magnitude of the design problem. Lens systems are made up of lenses having spherical surfaces, and such surfaces do not form perfect images. This defect may be largely counteracted by combining lens shapes and different glasses. The principal defects in the image are the result of chromatic aberration, spherical aberration, distortion, curvature of field, astigmatism, coma, and lateral colour.

Chromatic aberration

White light in passing through a prism is broken up into its constituent colours, the wave lengths of which are different. A simple or compound lens, composed of only one material, will exhibit different focal lengths for the various constituents of white light. This is due to the dispersive power of the lens. Every wave length is differently refracted, the shortest waves most and the longest waves least. The blue-violet rays cross the lens axis first and the red rays last. There will be a series of coloured foci of the various constituents of white light extending along the axis.

As a result, the lens will not produce a sharp image with white light. Instead, the image will be surrounded by coloured zones or halos which interfere with the visual observation of its true colour. This is spoken of a chromatic aberration. It may be lessened by reducing the aperture of the lens, or better still, by using a lens composed of more than one material (compound lens). Two or more different glasses or minerals are necessary for correcting the chromatic aberration of an objective, and the amount of correction depends upon the dispersive powers of the components of the objective. If two optical glasses are carefully selected to image light of two different wave lengths at the same focal point, the lens is said to be achromatic, and an objective containing such a lens system is spoken of as an achromatic objective. The remaining rays of the white light will be imaged at approximately the same point.

An achromatic objective will yield images free from pronounced colour halos. If the focus is shifted slightly, faint green and pink halos may be observed. The slight residual colour will not prove objectionable for the usual microscopic work. Achromats are the universal objectives for visual work and are very satisfactory in photomicrography when used in monochromatic light (obtained by the use of filters). Lens systems corrected for light of three different wave lengths are called apochromatic objectives. These objectives are composed of fluorite in combination with lenses of optical glass. The images produced by objectives in this group exhibit only a faint blue or yellow residual colour. Since these objectives are corrected for three colours instead of for two, they are superior to the achromats. Their finer colour correction makes possible a greater usable numerical aperture.

The violet rays are brought to the same focus as visual rays. This fact makes these objectives excellent for photographic use for both white and monochromatic light. Another group of objectives exhibit qualities intermediate between the acrhomats and the apochromats. They are called *semiapochromats*. If the mineral fluorite is used in their construction, they are termed fluorite objectives. These objectives also yield excellent results when used for photomicrography.

Spherical aberration

This refers to the greater power in the outer portion of a spherical surface than in the inner portion. This is overcome by judicious combinations of convergent and divergent lens elements, properly shaped to minimize the variation of focal power with aperture. Spherical aberration causes some of the light which should be in the central spot to diffuse out into the ring structure. This causes a loss in contrast in the normal microscope preparation.

Distortion

This type of aberration renders a square object as an image with curved sides. If the rulings near the edge appear curved inward, it is known as cushion distortion. If the opposite effect occurs, where the rulings appear curved outward, it is known as barrel distortion. Distortion is caused by the lens surface having different magnifications at the marginal and central portions of the image.

Curvature of field

This aberration is caused by a spherical lens surface which produces a curved image of a flat object due to the marginal portions of the image coming to a focus at a different distance than the central portions of the image

Astigmatism

If a marginal point object is drawn out into two separate line images lying at different distances from the lens surface, it is called astigmatism. It results in a general deterioration of the off-axis image. An astigmatic image can never be focused sharply except for detail parallel or perpendicular to a radius of the field.

Coma

This name is given to the defect in which different circular concentric zones of the lens surface give different magnifications to an off-axis image. This results in a point object being imaged as a comet-shaped image. Coma in the center of the field is an indication of damage to the objective.

Lateral colour

The presence of this defect results in light of one colour being imaged at a greater magnification than light of another colour. This causes an off-axis image of a point object to be spread out into a tiny spectrum or spread of colour.

Oculars

The chief functions of the ocular, or eyepiece, are the following:

1. It magnifies the real images of the object as formed by the objective;
2. It corrects some of the defects of the objective;
3. It images cross hairs, scales, or other objects located in the eyepiece.

Several types of eyepiece are employed, depending upon the kind of objectives located on the microscope. Those most commonly used are known as Huygenian, hyperplane, and compensating oculars.

Huygenian eyepiece

In this type of eyepiece, two simple plano-convex lenses are employed, one of which is below the image plane. The convex surfaces of both lenses face downward. Oculars in this group are sometimes spoken of as negative eyepieces. This type of ocular is made with a large field lens, which bends the pencils of light coming from the objective toward the axis without altering to any great extend the convergence or divergence of the rays in the individual pencils. Above the field lens, and at some distance from it, is a smaller lens known as the eye lens, the functions of which is to convert each pencil of light into a parallel or only slightly diverging ray system capable of being focused by the eye. The rays, after emerging through this lens, then pass through a small circular area known as the Ramsden disk, or eyepoint. It may be seen that the real image of the object is formed between to two eyepiece lenses. In an eyepiece of this type, the distance separating the two lenses is always a little greater than the focal length of the eye lens.

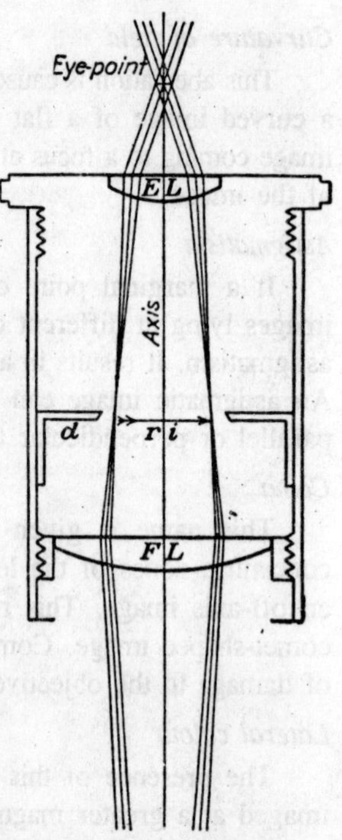

Fig. 18.9. Huygenian eyepiece.

The reason for this is to prevent any dirt on the field lens from being seen sharply focused by the eye. An image should be viewed with the eye placed at the Ramsden disk in order to obtain the largest field of view sand also to obtain the maximum brightness over the field. The Huygenian eyepiece works well with the low-power achromats but gives under-corrected curvature of field and lateral colour with the intermediate and higher power objectives. The degree of compen-sation required increases with the objective power, making it highly desirable to have a graded series of eyepieces. Therefore, the Huygenian eyepiece should be used to cover the low powers; the hyperplane eyepiece, the intermediate powers; and the compensating eyepiece, the high powers.

Hyperplane eyepiece

Apochromatic objectives, when used with compensating eyepieces, give fields that are not flat. Flat-field eyepieces have been designed to correct this defect. They give much flatter fields than do the other two types but they are lless perfectly corrected chromatically. Oculars of this type are referred to as hyperplane, planoscopic, periplane, etc. They may be employed with the higher power achromatic, fluorite, and apochromatic objectives without introducing chromatic aberrations in the image. Their colour com-pensation falls about midway between the Huygenian and the compensating eyepieces.

Compensating eyepiece

Oculars of this type consist of an achromatic triplet combination of lenses. These eyepieces are more perfectly corrected than are those of the Huygenian and hyperplane types. A compen-sating eyepiece is corrected to neutralize the chromatic difference of magni-fication of the apochromatic objectives. Such eyepieces are intended, therefore, to be used primarily with apochromatic objectives, although they may be employed with the higher power achromatic and fluorite objectives with good results.

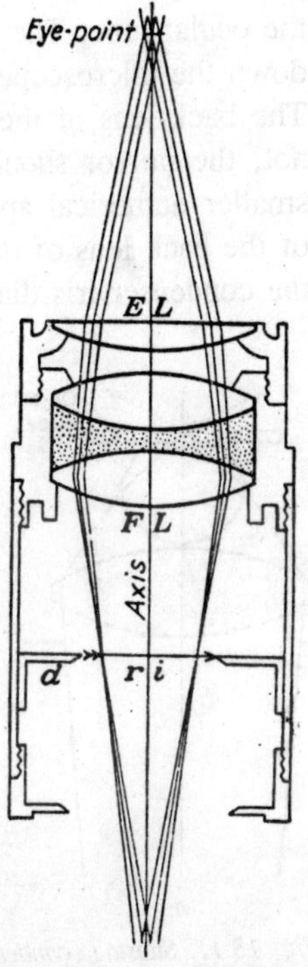

Fig. 18.10. Compensating eyepiece.

Condensers

Several methods are employed for illuminating the object under examination. In bacteriology, the two methods commonly used are (1) illu-mination by transmitted light and (2) dark-field illumination.

Illumination by transmitted light

A condenser may be defined as a series of lenses for illuminating, with transmitted light, an object to be studied on the stage of the microscope. It is located under the stage of the microscope between the mirror and the object, whereas the objective and ocular lenses are located above the stage. It is sometimes referred to as a substage condenser. The most popular substage optical system is known as the Abbe condenser. A condenser is necessary for the examination of an object with an oil-immersion objective to obtain adequate illumination.

A condenser is also preferable when working with high-power dry objectives. Probably the most commonly employed condenser has a 1.25 N.A. A good condenser sends light through the object under an angle sufficiently large to fill the aperture of the back lens of the objective. When this is accomplished, the objective will show its highest numerical aperture. This may be determined by first focusing the oil-immersion objective on the object. The eyepiece is then removed from the ocular tube. The back lens of the objective is observed by looking down the microscope tube, care being taken not to disturb the focus. The back lens of the objective should be evenly illuminated. If it in not, the mirror should be properly centered. If the condenser has a smaller numerical aperture than the objective, the peripheral portion of the back lens of the objective will not be illuminated, even though the condenser iris diaphragm is wide open.

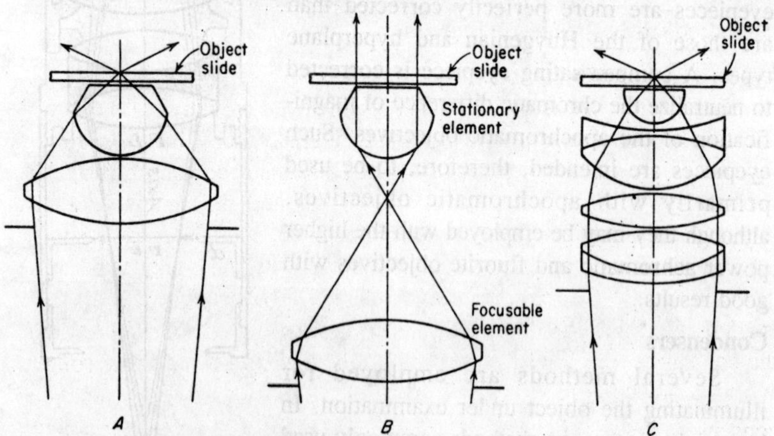

Fig. 18.11. Substage condensers. A—Abbe condenser; B—Variable-focus condenser; C—Achromatic condenser.

If the condenser has a greater numerical aperture than that of the objective, the back lens of the objective may receive too much light, resulting in a decrease in contrast. The smaller the aperture, the greater the depth of focus and the greater the contrast of the components of the image. The lowest permissible aperture is reached when diffraction bands become evident about the border of the object imaged. This difficulty may be largely overcome by closing the iris diaphragm of the condenser until the leaves of the iris appear around the edges of the back lens of the objective. The diaphragm is then said to be properly set. The setting of the iris diaphragm will vary with different objectives. The Abbe condenser is a 1.25- N. A. condenser utilizing only two lenses. Because of its simplicity and good light-gathering ability, it has become extensively used for general microscopy.

It is, of course, not corrected for spherical or chromatic aberration, but for general visual observation it serves very well. The variable-focus condenser is a two-lens condenser, 1.25 N. A. maximum, in which the upper lens element is fixed and the lower one focusable. By this means it is possible to fill the field of low-power objectives without the necessity of removing the top element. This condenser is basically similar to the 1.25 N. A. Abble when the lower lens is raised to its top position. When the focusable lens is lowered, the focus of the light is brought in between the elements, and when this focus is at the point indicated in the diagram, the light emerges as a large-diameter parallel bundle. The achromatic condenser is a 1.40- N.A. Condenser that is corrected for both chromatic and spherical aberrations. Because of its high degree of correction, it is recommended for research microscopy and colour photomicrography where the highest degree of perfection in the image is desired.

Dark-field illumination

The microscope is most commonly employed by allowing the light to pass through the object. This is called microscopy in transmitted light or bright-field microscopy. An object cannot be seen in bright-field microscopy unless it absorbed or refracts the light passing through it. Contrast is thus set up between the object and the surrounding medium. Objects that display feeble contrast with the background are difficult to see in bright- field illumination. If the aperture of the condenser is opened completely and a dark-field stop inserted below the condenser, the light rays reaching the object form a hollow cone.

If a stop of suitable size is selected, all the direct rays from the condenser can be made to pass outside the objective. Any object within

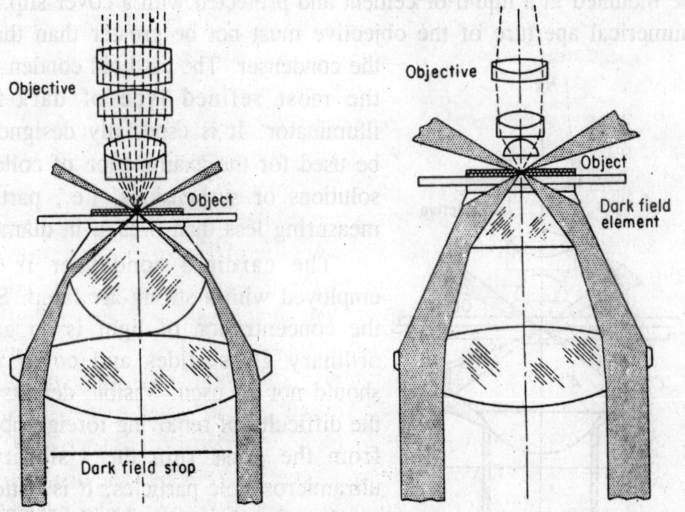

Fig. 18.12. A—Abbe condenser with dark-field stop inserted below the condenser. B—Abbe condenser.

this beam of light will reflect some light into the objective and be visible. This method of illuminating an object, where the object appears self-luminous against a dark field, is known as dark-field illumination. Three types of condensers are employed for dark-field illumination: (1) the Abbe, (2) the paraboloid, and (3) the cardioid.

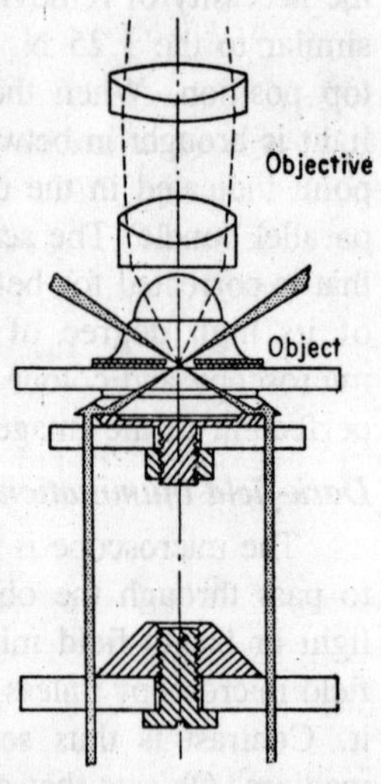

Fig. 18.13. Paraboloid condenser.

The Abbe condenser is probably more commonly employed than the other two because it is especially suitable for objects that do not require the highest magnifications to make them visible. It may be employed either by inserting a dark-field stop below the condenser or by unscrewing the top part of the condenser and substituting for it a dark-field element. The paraboloid condenser is designed to be used with high-power oil-immersion objectives and an intense source of light.

In using this condenser, it is necessary to place cedar oil or glycerin between the condenser an the slide. Also, the specimen must be mounted in a liquid or cement and protected with a cover slip. The numerical aperture of the objective must not be greater than that of the condenser. The cardioid condenser is the most refined type of dark-field illuminator. It is especially designed to be used for the examination of colloidal solutions or suspensions, i.e., particles measuring less than 0.25 μ in diameter.

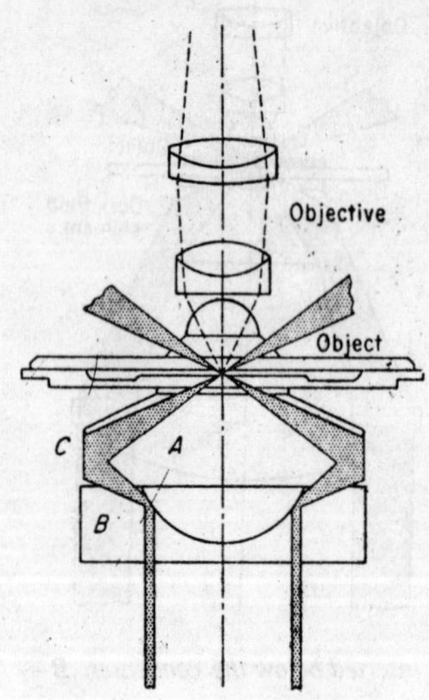

Fig. 18.14. Cardioid condenser.

The cardioid condenser is best employed with a strong are lamp. Since the concentration of light is so great, ordinary glass slides and cover slips should not be used. Visible defects and the difficulty of removing foreign objects from the glass ruin the visibility of ultramicroscopic particles. It is better to employ fused-quartz object slides and fused-quartz cover slips. They are free from bubbles and other imperfections and can be heated in a flame to drive off all dirt, after being chemically cleaned. Pijper (1951) recommended the use of thin microscope slides to which were cemented pieces of mica of similar shape by means of Canada balsam. When the balsam had hardened, a thin layer of mica was split off with a sharp knife, which left an untouched, scratch-free, and dust-free surface, perfectly suited for dark-field microscopy.

Phase-contrast Microscope

The introduction of phase-contract microscopy is probably the greatest single advance in biological laboratories in this century. For the first time, living organisms and cells may be examined in detail without previous treatment, and the image produced a true one. All living cells and organisms, although appearing almost homogeneous when examined unstained by ordinary microscopic methods, are composed of minute structures having slight differences in refractive

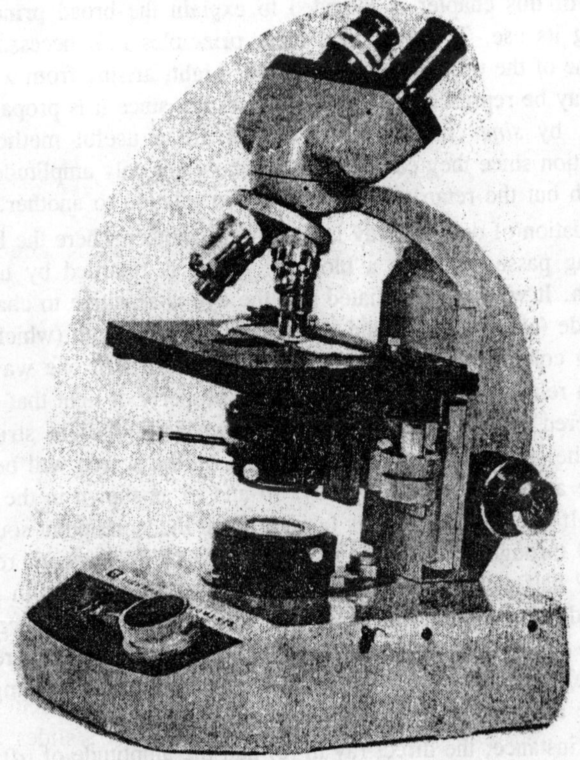

Fig. 18.15. The phase-contrast microscope.

index. The phase-contrast microscope, by converting these slight differences in refractive index into changes of amplitude (or brightness), produces an image that may be accurately focused and one that is easily seen or photographed.

Professor Zernicke, who was awarded the Nobel prize for his work on phase contrast, first applied his original work on telescopes to the biological microscope in 1935, but it was not until 1945 that a commercial model was available in Great Britain, although Burch and Stocks described a method for the conversion of an ordinary microscope in 1942. These microscopes are now being widely used, and there seems little doubt that they will in the future replace what we now know as the routine microscope, since they may be used either for phase-contrast microscopy or for routine purposes.

Principles of Phase-contrast Microscopy

Without going too deeply into the theory of phase contrast, the first part of this chapter is intended to explain the broad principles underlying its use. To understand these principles it is necessary to recall some of the properties of light rays. Light, arising from a point source, may be represented by straight lines or, since it is propagated in waves, by *sine* curves. These curves are a useful method of representation since they can be made to show not only amplitude and wavelength but the retardation of one ray in relation to another.

Retardation of one light ray in relation to another where the lower ray, having passed through a block of glass, is retarded by half a wavelength. It will be appreciated that the eye is sensitive to changes in amplitude (or brightness) and to changes in wavelength (which are changes in colour) but not to changes in phase, where one wave is retarded in relation to another. One further property of light that must be considered is that of interference. If two rays of light strike a screen at the same point, the resultant light on the screen will be the sum of the amplitude of the two rays (*b* and *c*) as shown in the *sine* curve (*d*). If one of the two rays (arising from the same point source) had passed through a block of glass, of an exact thickness to retard that ray by half a wavelength, instead of getting an increase in light there would be no light. This is because coherent light rays have the property in interfering with each other. This interference may result in complete extinction of the light or in a reduction of it depending on the relative amplitudes of the light rays.

If, for instance, the direct ray in (*e*) had the amplitude of (*d*) and the retarded ray the amplitude of (*b*) instead of extinction of the light,

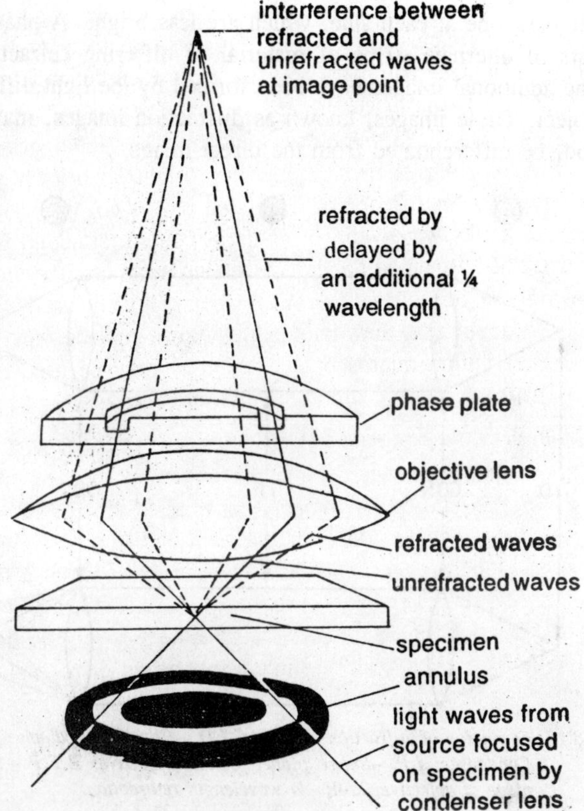

Fig. 18.16. Path of light rays in a phase contrast light microscope.

there would have been light of the amplitude of (*c*). For practical purposes it may be said that when coherent light rays interfere the amplitude of the resultant ray can be obtained by subtracting the amplitude of one from the other. Fractional phase differences (for example, ¾ or ¼ of a wavelength) between rays will result in partial interference and in this way an image of an unstained object may be built up.

By almost closing the iris diaphragm of the substage condenser of the normal microscope an image of this small aperture is formed at the back focal plane of the objective. This image, which can be seen by removing the eyepiece, may be focused with an auxiliary eyepiece. If a phase grating is placed in the object plane (on the stage), instead of only one image being seen at the back focal plane of the objective three images are visible, the original bright one in the centre, and a further two, one at each side, which are less bright. A phase grating consists of alternate strips of material of differing refractive index and the additional images have been formed by the light diffracted by the object. These images, known as diffraction images, may, by this method, be differentiated from the direct image.

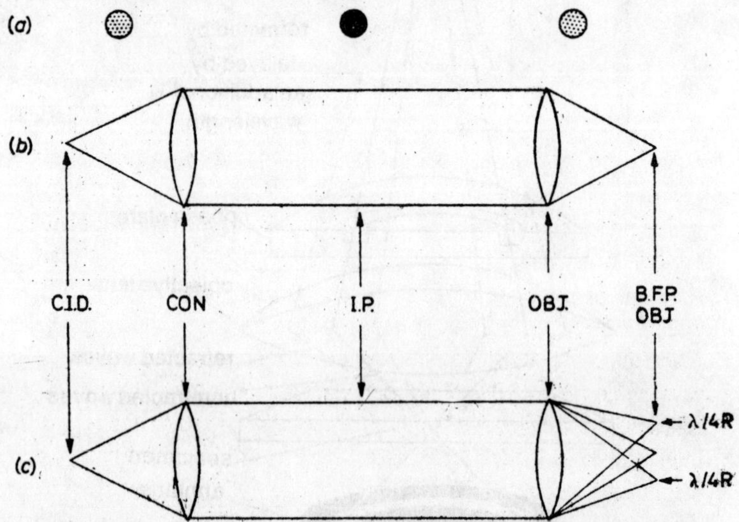

Fig. 18.17. Formation of diffraction images. C.I.D.—Closeed iris diaphragm; CON—Condenser; I.P.—Image plane; OBJ.—Objective; B.F.P.—Black focal plane of objective; λ/4R—¼ wavelength retardation.

One other effect that has taken place, which is not apparent but can be proved, is that the diffraction images are out of phase with the direct rays by ¼ of a wavelength (¼ λ), that is, they have been retarded by a ¼ λ in relation to the direct rays. If an annulus is placed in the substage condenser an image of that annulus will be formed in the back focal plane of the objective, and an object possessing slight non-homogeneities (such as unstained living cells) placed in the object plane will produce a halo of light both inside and outside the annular image. This halo is composed of light rays which have been diffracted by the object and are ¼ λ out of phase with the direct light rays. It will follow that if the diffracted rays of light could be retarded a further ¼ λ, then the phase difference between the direct and diffracted rays would be ½ λ and interference would take place in the final image plane, building up a picture in light and shade, of an unstained specimen. Zernicke devised the 'Z' plate, now known as the

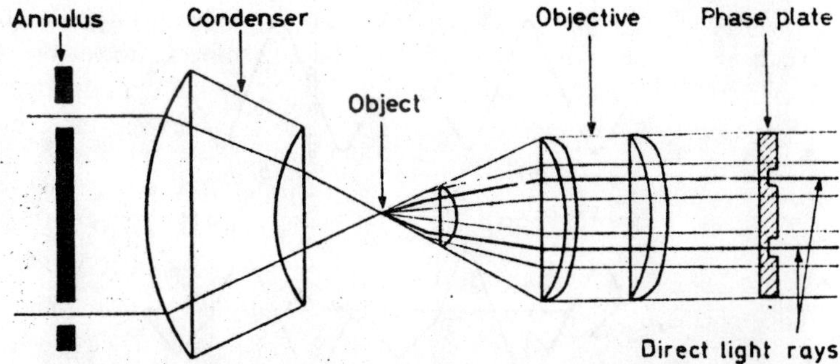

Fig. 18.18. Passage of light rays through the optical components of the phase-contrast microscope.

phase plate, which, placed at the back focal plane of the objective, brought this about.

The phase plate consists of an optically plane glass disc out of which is cut a channel to coincide with the image of the light annulus. The depth of the channel must be the exact depth to retard the diffracted rays, which travel through the full thickness of the plate, by a ¼ λ in relation to the direct rays which travel through the channel. Although interference will now take place, the great difference in amplitude (or brightness) between the two sets of rays will prevent the maximum contrast from being obtained. To overcome this factor light-absorbing material is deposited in the area of the channel which reduces the amplitude of the direct light without affecting the diffracted light, this permitting the maximum contrast to be obtained.

A broad summary of these principles is (*a*) a ray of direct light from the annulus, on passing through the object (*b*) gives rise to a diffracted ray (dotted line), which is retarded by ¼ λ; (*c*) on passing through the phase plate the diffracted ray, retarded by a further ¼ λ is now in a position to interfere with the direct light ray; (*d*) the amplitude of the direct light ray is reduced after passing through the light-absorbing material and better contrast is obtained. Although in the illustration, for the sake of clarification, (*c*) and (*d*) take place separately, in practice they occur almost simultaneously, and interference, of course, does not take place until the images are once again combined at the real image plane in the eyepiece. The foregoing theory should only be true if monochromatic light is used as an illuminant, since white light would be split into its component colours when diffracted; in practice, however, white light may be used but

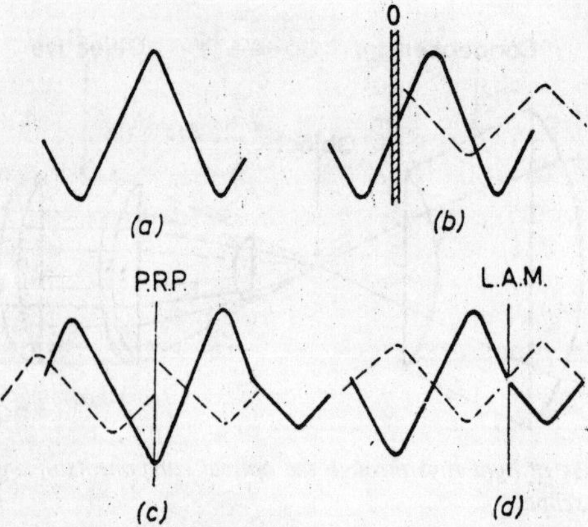

Fig. 18.19. Diagrammatic summary of tne theory of phase-contrast microscopy. O—Object; P.R.P.—Phase retarding plate; L.A.M.—Light absorbin material.

better contrast is obtained by using a mercury green filter in conjunction with a compound high intensity lamp.

Equipment

Phase-contrast equipment need not be very expensive and several papers have been written describing methods of adapting normal microscopes for phase contrast at a small cost. The performance of such a converted microscope is not usually equal to a specially designed commercial model, but will suffice for routine purposes. The microscope described in this chapter will be that available from Vickers Ltd.

There are several other models available commercially which give equally good results by using a cross-shaped or slit-shaped aperture instead of an annulus in the substage, but since the underlying principle is the same they are no described.

Lamp

An intense source of illumination should be used, such as a high intensity compound lamp, with a mercury green (Wratten 62) filter.

Annulus

A different sized annulus will be required for each objective. These may be inserted separately, but a rotary changer carrying a set of annuli, one for each objective, mounted below a special condenser is more convenient; each of these annuli can be centred by means of two centring screws.

Objectives

Objectives are supplied with phase plates already fitted, and since the phase plates affect their performance only slightly when used in a normal manner, they may be used without an annulus for routine microscopy.

Auxiliary Microscope

The auxiliary microscope is used for examining the back focal plane of the objective and ensuring that the objective phase plate and the condenser annulus are properly adjusted.

Setting up the Microscope

1. The microscope is set up in the usual way, ensuring that there is no annulus in the substage.
2. Focus on the object, closing the iris diaphragm if necessary.
3. Rotate the annulus changer until the appropriate annulus is in position. Without disturbing the focus remove the eyepiece and replace it with the auxiliary microscope.
4. Adjust the auxiliary microscope to bring the image of the phase plate into sharp focus.
5. If the image of the light annulus does not coincide with the grey ring of the phase plate, it is adjusted with the centring screws until its image is concentric with, and completely covered by, the grey ring of the phase plate; the condenser may need to be raised or lowered slightly to adjust the size of the image of the light annulus. If the light annulus is not evenly illuminated the light should be adjusted.

Note. This procedure should be repeated each time the objectives is changed.

Fluorescence Microscope

Fluorescence is a property of some substances to excite emission of visible light by the absorption of invisible ultraviolet radiations. The exact mechanism of this phenomenon is not clearly understood. Some materials are autofluorescent, whereas others can be made to fluoresce by treatment with fluorescent chemicals called fluorochromes. Equipment includes a source of ultraviolet rays, suitable filters for isolating the radiations, and a standard microscope. Some applications of fluorescence microscopy include examination of acid-fast bacteria (*M. tuberculosis*, etc.), differentiation of living and dead micro-organisms, examination of hair for mold spores, and location of

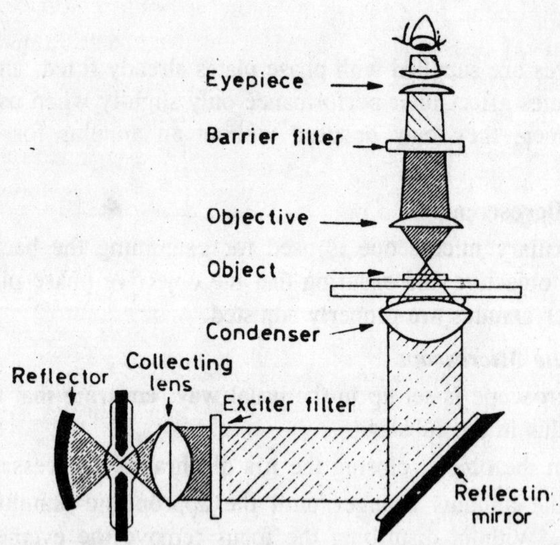

Fig. 18.20. Diagram to illustrate the component parts of the fluorescent microscope.

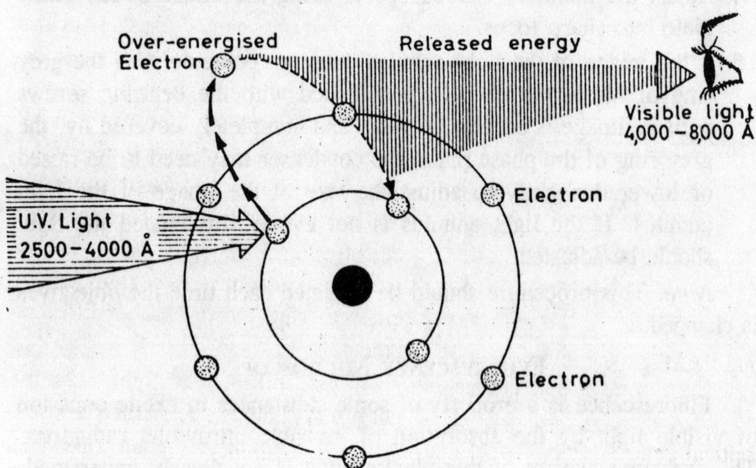

Fig. 18.21. Diagram to illustrate how ultraviolet light may excite fluorescence in a molecule.

Polarizing Microscope

Although it is beyond the scope of this book to enter deeply into the theory of polarized light, the basic elements of the subject will be explained. Light is assumed to be due to a wave motion, to the upward and downward vibration of ether particles. These do not move along in the direction of the light ray, but vibrate at right angles to it, and when the light ray ceases they return to their original position. Ether is supposedly an homogeneous medium and there is no reason therefore to believe that these particles will vibrate in any one direction more than another. To explain polarized light it is necessary to suppose that light normally vibrates in all planes.

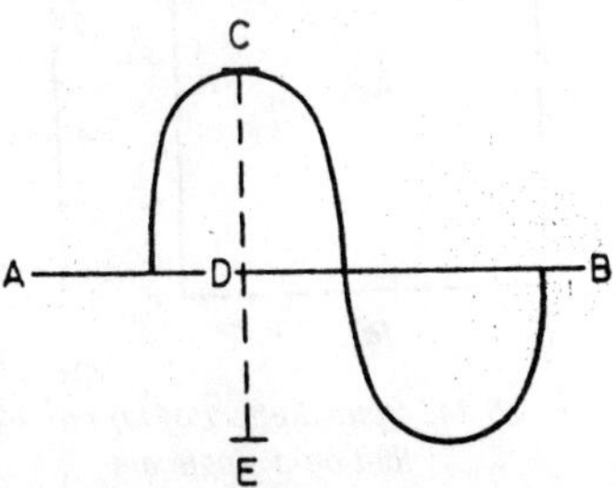

Fig. 18.22. Sine curve representing a wave of light. AB—direction of travel; CDE—direction of vibration.

It is difficult to imagine a particle oscillating in all planes at one time, but it is possible to imagine that it moves—at right angles to the direction of the light ray—in all planes in such rapid succession so as to act as if it were moving in all these planes at one time. While this theory is not strictly accurate, it is sufficiently correct to explain the behaviour of polarized light. If a dot drawn on a sheet of white card is viewed through a block of glass laid on top, only one dot will be visible from above. If the block of glass is replaced by a polished block of crystal, such as Iceland spar, two dots will be visible. Such a crystal is described as being bi-refringent or anisotropic; it has split each light ray from the dot into two rays which emerge from the crystal at different points.

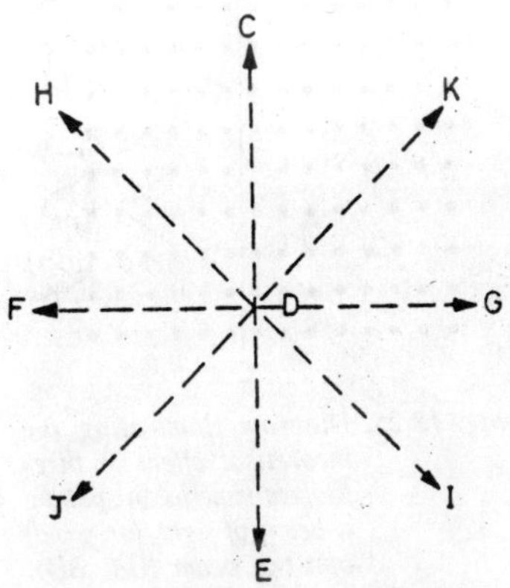

Fig. 18.23. Theoretical representation of the vibration of light in all planes.

This splitting of light rays by certain crystals is due to their uneven optical density. It is known that light rays are retarded when travelling through an optically dense medium such as glass, but since the molecules in glass are evenly spaced in all directions only a simple retardation or slowing takes place. The molecular structure of a crystal differs from glass in that although its molecules are regularly spaced they are closer together in one direction than in another; they are therefore unevenly dense. There are many types of crystalline structure, but all have the common property of being more dense in one direction than in another.

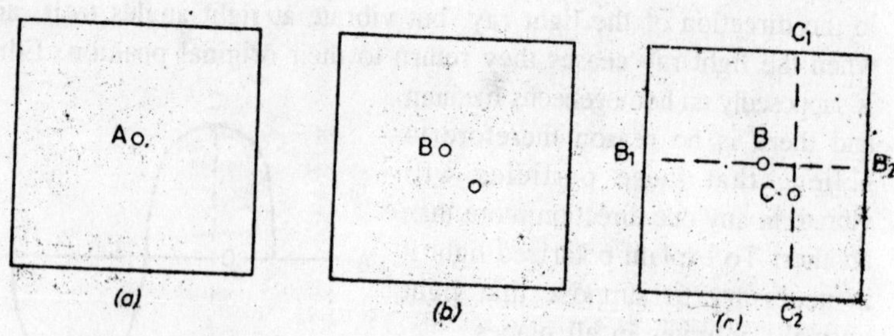

Fig. 18.24. Optical effect of (a) block of glass and (b and c) birefringent material when laid on a single dot.

Figure 18.25 shows a downward view of a series of posts through which wind is blowing from directions W1 and W2. It will follow that from whatever angle the wind is blowing it can only leave in the direction of A or B. The intensity of the wind emerging at points A or B will depend on the angle at which it enters. If it enters from W1 then almost all the wind will emerge in direction B; if the wind enters from W2, an almost equal amount will emerge in each direction. If we now substitute the words 'crystalline structure' for 'posts driven into the ground', and 'light rays' for 'wind', an understanding may be gained of what happens when a light ray passes through a crystal: the ether particles are vibrating in all directions at right angles to the line of propagation when it enters, but two rays emerge, and each of these causes ether particles to vibrate in one plane only.

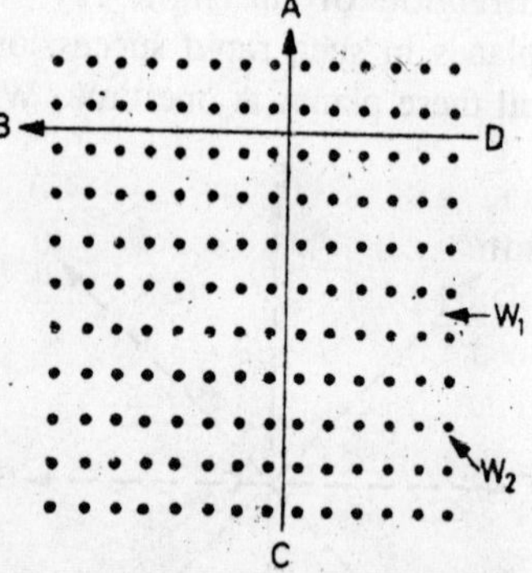

Fig. 18.25. Diagram illustrating the theoretical effect of birefringent material in splitting a beam of light (or wind) into two beam (CA, BD).

Light when entering a dense medium is retarded in speed. Further, being an unevenly dense medium, the crystal will retard the two rays to a differing degree, and since refraction is partly dependent on density, the two rays will be refracted or bent to differing degrees. This is known as double refraction or birefringence and explains the phenomena described above. A ray of light entering such a crystal will be converted into two rays which will emerge at different points, and the emergent light rays will be polarized; that is, all the vibrations in one ray (B) will be in one single direction (B_1–B_2); in the other ray (C) in another single direction (C_1–C_2), and these directions will be at right angles to each other.

Nicol Prism

Just over 100 years ago, Nicol devised a prism from which light rays, having passed through, would emerge vibrating in a single plane, that is, as polarized light. The single direction in which the light is vibrating when it emerges is known as the 'optical path' of the prism. The prism is composed of a crystal of Iceland spar, cut to the shape slit in half and the halves cemented together with Canada balsam along the line CB–CB. On entering the prism, a light ray (A) is divided into two rays (B and C) which are refracted differently, ray C being refracted to one side.

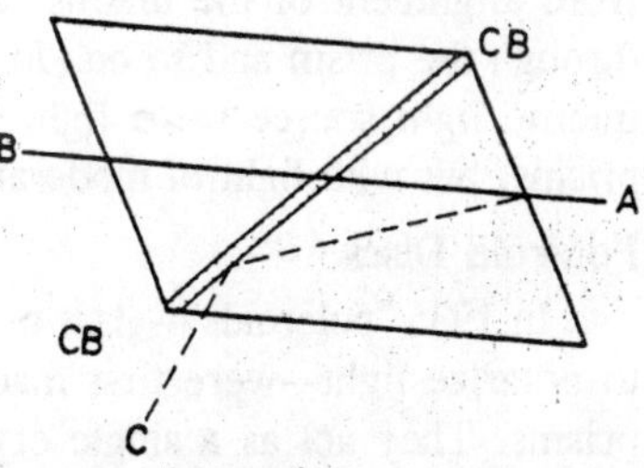

Fig. 18.26. The Nicol prism.

Owing the difference in the refractive index between Canada balsam and the calcite spar crystal and the cement, ray C on meeting the surface CB–CB at a greater angle than ray B, is totally reflected out of the prism. Ray B passes through the prism and emerges vibrating in the direction of the optical path of the prism only and is polarized light. It will follow that if another Nicol prism is placed above the first one, the polarized light ray B will pass through the upper prism if their optical paths are aligned (B_1) but if the upper prism is rotated through 90 degrees so that the optical paths of the prisms are crossed, then ray B will be totally reflected out of the upper prism (B_2). Such prisms are said to be crossed and it will be seen that no light will normally emerge from crossed Nicol prisms. If the upper prism is slowly rotated it will be seen that the amount of light passing through will vary with their relative positions. At a rotation of 45 degrees,

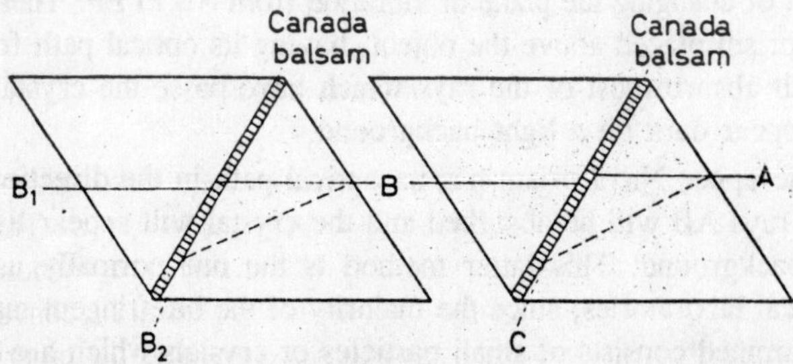

Fig. 18.27. The effects of two Nicol prisms, with optical paths aligned (ABB_1) and optical paths crossed (ABB_2).

from alignment of the prisms, approximately half the light will pass through the prism and so on. In practice, it will be found that with an intense light source some light will pass even through crossed Nicol prisms, but with light of moderate intensity the field will appear black.

Polaroid Discs

In 1935 'polaroids'—glass or celluloid covered discs with the ability to polarize light—were first made available for use in place of Nicol prisms. They act as a single crystal of herapathite which is not only birefringent, but has the ability to absorb the ordinary ray (which would be refracted out of Nicol prism, only the extraordinary ray being transmitted. Polaroids are made by suspending ultra-microscopic crystals of herapathite in nitrocellulose. All the crystals in the suspension are orientated so that their optical paths are aligned. This suspension when mounted between two glass plates or celluloid sheets acts as a single crystal. One glass plate is made to fit into the substage filter carrier, and the other has a metal mount to hold it in place on top of the eyepiece. The celluloid sheet may be cut with scissors and used in a similar manner. For all practical purposes they may be used as Nicol prisms.

Use of the Polarizing Microscope

It has been shown that certain crystals have the power to convert a single ray into two rays of light, which are vibrating in a single plane at right angles to each other, and also have the power of quenching or absorbing one set of these rays. If such a crystal is placed on the stage of a microscope having a Nicol prism (or polaroid) in the substage. The light rays in the filed will be vibrating in the optical path of the Nicol prism (AB), except those that pass through the crystal, which will be vibrating from C to D or E to F. If the set of rays CD were absorbed (as described above for herapathite), the crystal would have the effect of changing the plane of vibration from AB to EF. Therefore, a Nicol prism placed above the object, having its optical path from A to B, will absorb most of the rays which have passe the crystal, and it will appear dark on a light background.

If the upper Nicol prism has its optical path in the direction EF, then the rays AB will be absorbed and the crystal will appear light on a dark background. This latter method is the one normally used in histological laboratories, since the majority of the birefringent material to be examined consists of small particles or crystals which are easier to see as light objects on dark background. It will be seen that such crystals have the apparent power of changing the plane of vibration

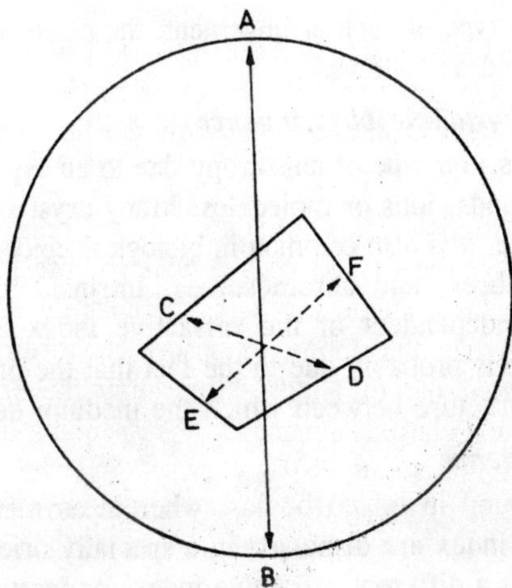

Fig. 18.28. Showing birefringent effect of a crystal.

(from AB to EF) and for this reason they have occasionally been referred to as 'optically active'. The direction in which the plane of vibration is changed, and the degree to which it is turned, may be used to assist in the identification of a crystal.

For this reason, petrological microscopes have the polarizing prism (in the substage) in a graduated circular mount which may be rotated through 360 degrees. In addition, the stage is usually graduated and may be rotated in the same manner, so that the object may also be orientated. Crystals which change the plane of vibration clockwise are called 'dextra-rotary', and anti-clockwise 'laevo-rotary'. In the histological laboratory, one is usually only concerned with simple birefringence. The microscope is set up, polarized discs normally being used. When the material being investigated is in the field, the upper polaroid (analyser) is rotated. The material will appear light on a dark background if it is birefringent or anisotropic. Talc crystals, hair, myelin, silica and collagen fibres are among the many birefringent substances found in histological sections.

Types of Birefringence

Certain crystals or tissue structures show more than one index of refraction for a given wavelength of light and are therefore said to be doubly refracting or birefringent. As was shown above, this is due to some sort of asymetrical and orientated spatial arrangement of particles. These particles carry resonating charges capable of interacting with the oscillations of light waves. Birefringent material may show one or more than one type of such arrangement; the more common types may be characterized as follows:

Intrinsic or crystalline birefringence

This refers to a type of anisotropy due to an asymetrical alignment of chemical bonds, ions or molecules. Many crystals display this type of birefringence, it is also common in biological objects such as collagen and muscle fibres, and chromosomes. Intrinsic birefringence in a specimen is independent of the refractive index of the immersion medium which is probably due to the fact that the orientated elements are of close structure between which the medium does not penetrate.

Form birefringence

This is found in mixed bodies, wherein asymetrical particles of one refractive index are dispersed in a specially oriented manner in a medium having a different refractive index. At least one dimension of the particles must be small in relation to the wavelength of light employed. These dispersed particles may be filaments, sheets, and so on, and they may be dispersed in a liquid, gas or solid; they can give rise to birefringence even if separately either or both are isotropic. Tests for form birefringence depend upon causing media of varying refractive index to penetrate between the particles when, at the appropriate R.I., form birefringence will disappear. (Examining objects mounted in a variety of mountants with differing R. I., for example, water, glycerol, HSR, and so on.)

Strain birefringence

When a dielectric substance is subjected to mechanical stress, the bonds within the substance can be distorted and give rise to a pattern which will result in birefringence. This is most simply demonstrated by twisting clear plastic (Perspex) between crossed polaroids when a birefringent spectrum of colour is produced. Similarly, glass or elastic tissue fibres under stress show birefringence.

Positive and negative birefringence

An object that appears bright on a dark field, when viewed between crossed Nicol prisms (or polaroids), is said to be birefringent (*see above*). This does not allow the determination of direction of the fast and slow axes of the doubly refracting material; having two different R.I.s one light ray will be retarded (slow) in relation to some distinguishing dimension of the object. In a collagen fibre the slow axis of transmission is parallel to the long axis of the fibre, the fibre is thus said to show positive birefringence with respect to its long axis; conversely, a chromosome shows negative birefringence with respect to its long axis.

To determine the sign of birefringence in a fibre or crystal (for example, to differentiate between the urate crystals of gout from calcium pyrophosphate dihydrate (CPPD) one needs, in addition to a pair of Nicol prisms (or polaroids), a first order red quarter-wave plate. Focus the object to be examined between crossed polaroids (or Nicol prisms) and insert the quarter-wave plate between the object and the analyser (upper prism) and rotate the plate until its slow axis of transmission (usually indicated by a stamped line or arrow) is parallel with the fibre or long axis of the crystals. The field will now be uniformly red (from its interference effect) and if the fibre or crystal appears blue then the slow axes of the plate and object are parallel and this is positive birefringence, if the object appears yellow it indicates negative birefringence (the slow axis of the plate is parallel with the fast axis of the object). If the slow axis of the quarter-wave plate is parallel with E—F and the crystal appears blue on a red background this is positive birefringence, if it appears yellow on a red background then it displays negative birefringence (with respect to its long axis). Dichroism is also detected by the use of a polarizing microscope.

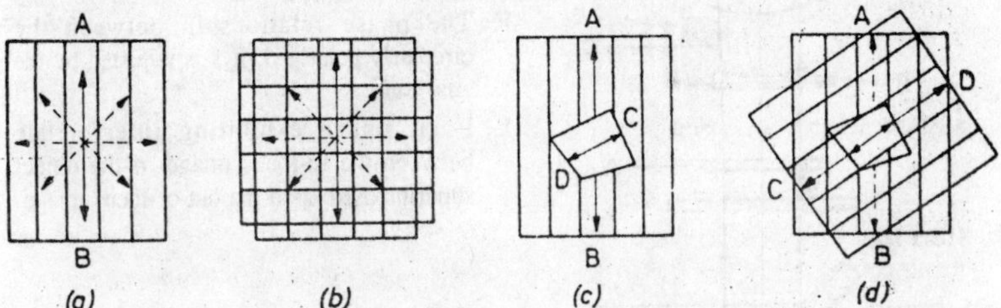

Fig. 18.29. Diagrammatic summary of the theory of polarization.

INTERFERENCE MICROSCOPE

Although a form of interference microscope has been in use for some years, it is only since World War II that it has been developed for use in the biological laboratory with increased sensitivity. It will now detect and accurately measure phase changes in an object of as little as 1/300 of a wavelength. There are two types currently in use in Great Britain; the Baker interference microscope, and the Dyson interference microscope. Since these microscopes are based on similar principles, only the Baker model will be discussed. The basic difference between the interference microscope and the phase-contrast microscope is that the former does not rely on diffraction by the object for

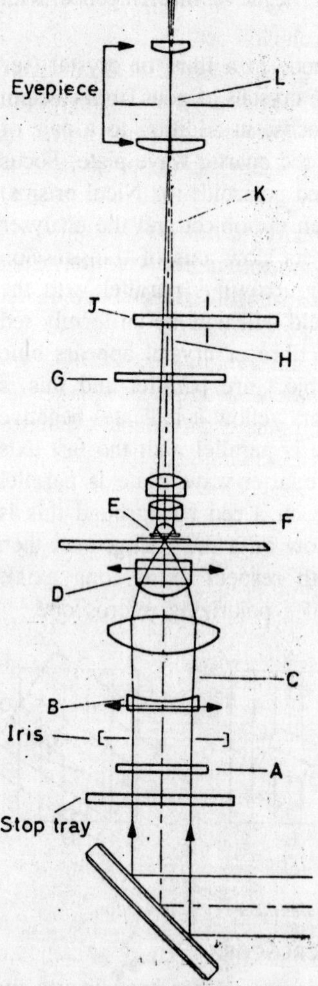

A. Swing-out polarizer. The rotation of this element controls the intensity relationship between the double-refracted beams, permitting the out-of-focus image to be extinguished for normal transmitted light conditions.
B. Double-refracting negative lens.
C. The double-refracted rays entering the Abbe condenser.
D. Double-refracting plate cemented to the front lens of the condenser, rendering it bi-focal.
E. Double-refracting plate rendering the objective bi-focal.
F. The re-combined double-refracted rays.
G. Quarter-wave plate.
H. The re-combined rays circularly polarized in opposite directions by the quarter-wave plate.
J. Rotatable analyser, with swing-out section, calibrated in degrees.
K. The phase relationship between the circularly polarized rays is adjusted by the analyser.
L. Final image exhibiting interference between the in-focus image of the object superimposed upon the out-of-focus image.

Fig. 18.30. Simplified diagram of the double refraction interference microscope for × 100 double-focus system.

interference, but generates mutually interfering beams which produce the contrast. It is this feature which enables such small phase changes to be seen and measured.

The two rays, which eventually combine to produce the final image, are formed by a plate of birefringent material immediately above the condenser. These two rays, having passed through the condenser, are re-combined by a similar plate of birefringent material at the face of the objective. Both these rays will have arisen from the same point of the light source, which is essential if interference is to take place in the final image, but the first will pass through one point in the image and the other through an area adjacent to it (the reference or comparison area). Consequently, each point in the final image is a compound one made up of two superimposed mutually different views of the same point of the object.

By using polarized light and an analyser the phase relationship between the two rays can be adjusted and measured. The goniometer analyser is calibrated in degrees, and a magnifier fixed above enables small movements to be read accurately. If white light is used as an illuminant, the various phases appear as different colours which change as the goniometer is rotated. Monochromatic light produces an amplitude contrast image and by rotation of the goniometer the degree of contrast can be varied for different components.

Uses

The microscope can be used for two purposes:

1. As an infinitely variable phase-contrast microscope with which individual parts of living cells may be studied with maximum detail; for this purpose there is little to be gained over the use of a good phase-contrast microscope, particularly since the condensers and objectives are matched and a change of objective also means changing and adjusting the condenser. A further disadvantage is that the high power (× 100) objective is a water immersion lens with the consequent limitation of resolution, and so on.
2. As a highly accurate optical balance, it may be used for estimating dry mass down to 1×10^{-14} g. The discovery that an increase in refractive index of 0.0018 is due to a 1 per cent increase in concen-tration of the solid substances contained in cells, and the fact that the refractive index of cell components can be estimated from the phase difference between them and the reference area (usually the fluid in which they are suspended) has made this possible. It is this aspect of interference microscopy that is at present being developed by many research workers.

Construction

Figure 18.30 is a diagram of the component parts of the Baker interference microscope and it will be seen that the instrument is an extremely complicated one, the use and adjustment of which requires great deal of practice.

ELECTRON MICROSCOPY

As already mentioned, particles less than half the wave length of the light employed in the microscope system cannot be resolved. This barrier to the microscopic examination of the majority of the viruses has been overcome by the use of the electron microscope. This instrument, developed over the past 20 years, employs electrons, which have a wave length of 0.05 Å, instead of light. The theoretic limit of resolution of the electron microscope is there fore very low, but in practice, commercial models give resolution of 10 Å or better. Although the simpler technics of electron microscopy have become standardized, considerable experience is still required before operators can obtain consistently good results from commercial instruments. Machines manufactured by the Radio Corporation of America, Siemens, and Philips are those best known to workers in North America and Europe. In general terms, electrons are generated from a tungsten filament

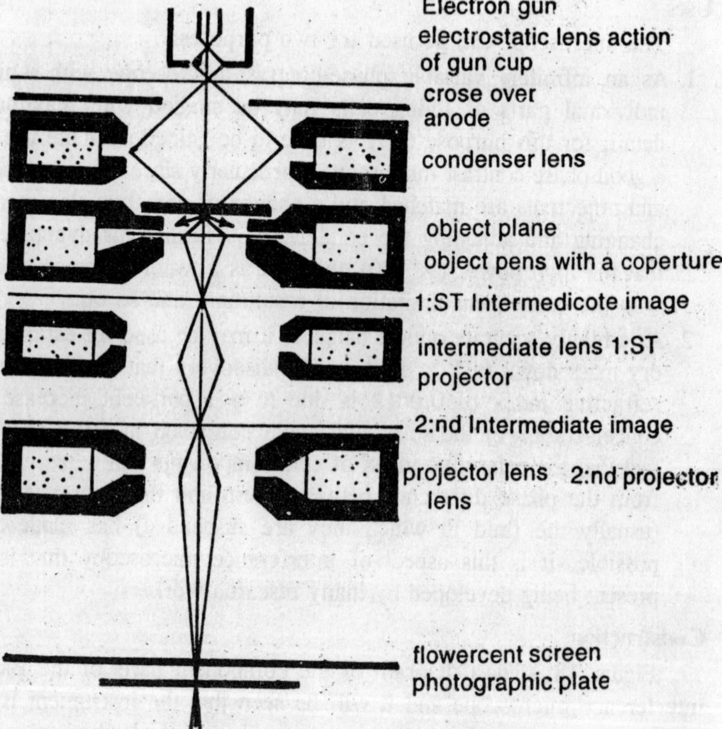

Fig. 18.31. The optical system of an electron microscope.

and pass down the microscope tube, which is kept evacuated. These electrons are next condensed by an electromagnetic lens and pass through the specimen.

In the specimen, dense material scatters or absorbs the electrons, which otherwise continue to pass down the tube. The electrons are magnified by the objective lens, and further by the intermediate and projector lenses, and finally reach a photographic plate or fluorescent screen, Direct magnifications of up to 100,000 are obtained. Specimens for electron microscopy are prepared in many ways. Many of the classic studies were done with preparations dried in air one collodion mounts. This process leads to some distortion of the shape of the virus particles. This has been largely replaced by a method in which the virus containing suspension is sprayed onto a carbon coated disk at the temperature of liquid nitrogen.

The disk is placed in *vacuo*, and the ice sublimes when the temperature is raised. In this freeze drying method there is much less distortion, and polyhedral instead of spherical forms are seen. The shadow-casting of air or freeze dried preparations is almost routine. A, thin layer of chromium, gold, platinum or uranium is deposited at a tangent so that the virus elementary bodies appear as it were in

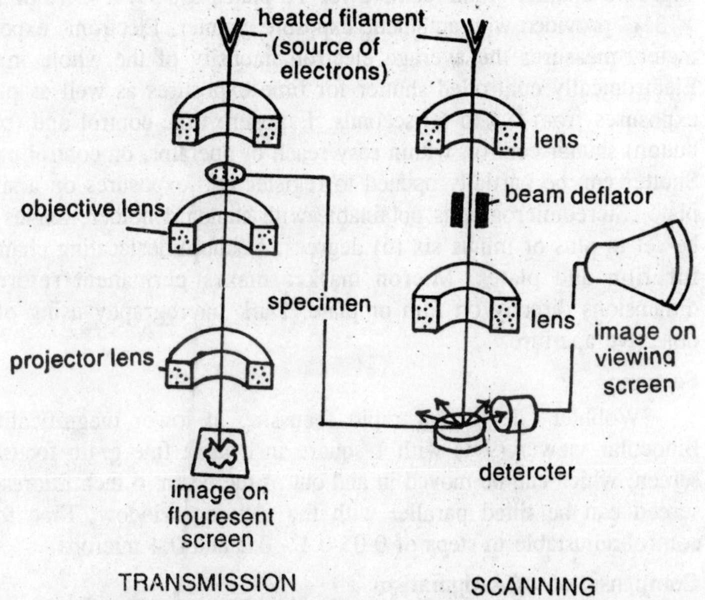

Fig. 18.32. Transmission and scanning electron microscope.

relief. Many electron micrographs of elementary bodies in this text are of shadowcast preparations. For purpose of measuring and counting elementary bodies, special technics have been devised, in which latex particles of known dimensions are mixed with the virus.

Resolving Power

10Å and better.

Magnification Range

Stepwise control, with the aid of a single panel control knob, without change of pole pieces, in 20 calibrated steps, each having a fixed magnification. 1,500-200,000 × on photographic plates. 700–90,000 × on 35 mm, film. Continuous control—which can be switched on at any time—by using several combinations of lenses, without change of pole pieces and specimen holder—from 300 × (for screening to 500, 000 × on 6½ in. Fluorescent screen.

Accelerating Voltage

40, 60, 80 and 100 KV, variable during operation.

Micrographic Facilities

35 mm. Film camera (40 exposures), provided with automatic exposure counter. Plate camera for 12 plates of, 3¼" × 4" or 3¼" × 3¼" provided with automatic exposure counter. Electronic exposure meter, measures the average electron intensity of the whole image. Electronically controlled shutter for time exposures as well as preset exposures from 0.5 to 60 seconds. Exposure time control and (push-button) shutter control, within easy reach of operator, on control panel. Shutter can be partially opened to register two exposures on a single plate. Stereomicrographs obtainable with standard holder, angles can be set to plus or minus six (6) degrees. Vacuum desiccating chamber for film and plates, Micron marker makes permanent reference dimensions directly on film or plate. Dark micrography using offset objective aperture.

Focussing

"Wobbler" device for rapid focussing at lower magnifications. Binocular viewer (9×) with 1 square inch ultra fine grain focussing screen, which can be moved in and out of the beam. 6 inch fluorescent screen can be tilted parallel with the viewing window, Fine focus control adjustable in steps of 0.05-0.1 - 0.2 and 0.4 microns.

Compensation of Astigmatism

Electromagnetic compensation of astigmatism on both the objective and 2nd condenser lens. Electromagnetic stigmators are centerable; no image shift during compensation. Compensation of objective lens astigmatism to less than 0.1 micron. Compensation is possible on any specimen, simultaneously with focussing, at magnifications from 20,000 × upward.

Electron Diffraction

Transmission diffraction patters of any party of the specimen obtainable within 30 seconds, without displacing the specimen. Diaphragm, consisting of four externally adjustable platinum blades, for obtaining diffraction patterns from areas as small as 1 square micron. Holder for three standard apertures, interchangeable with normal diffraction diaphragm holder, for high-selective diffraction studies. Central beam stop.

Specimen Chamber Cooling

Norelco cooling device lowers temperature of specimen chamber to approximately —90 degrees centigrade, preventing specimen contamination and decreasing contamination of aperture.

Specimen Holder

Single specimen holder for normal work, for the entire magnification range, for stereomicrographs and for electron diffraction. Specimen holder for 3 mm, specimen grids.

Aperturing

Multiple aperture holders of similar type for first and second condenser lens, for objective lens and for diffraction lens (high resolution diffraction). Three standard disk-type apertures in each holder. Aperture holders can be externally adjusted and centered. Wide range (from 25 to 300 microns) of apertures available.

Electron Optical System (Water Cooled)

V-shaped tungsten filament with D. C. Supply, external stepwise emission control. Double condenser lens, providing minimum beam spot of 1-2 microns diameter on specimen; condenser system provided with two aperture holders, electro-magnetic stigmator and beam deflection device; change from single to double condenser lens operation and *vice versa* within 30 seconds. Objective lens provided with electro-magnetic stigmator and aperture holder and "wobbler" device. Specimen stage, with smooth and drift free spacimen translation, permitting entire bridged area specimen to be scanned. Specimen is placed between the objective lens pole pieces (Norelco's exclusive "immersion" lens). Space between pole pieces equals 5 mm. Diffraction lens (for the lower range of magnifications) provided with externally adjustable diffraction diaphragm or standard aperture holder for high-selective diffraction studies; stepwise as well as continuous current control. Intermediate lens for the upper range of magnifications; stepwise as well as continuous current control. Projector lens having continuous current control as well as fixed magnification for fully illuminated screen. Projection chamber, provided with micron marker, automatic shutter, film camera intermediate focussing screen (tiltable parallel to the viewing window 6½ in, fluorescent screen, binocular viewer (9×) and plate camera.

Stabilization

Electrical stability: 1:200,000 for lens currents. 1:100,000 for the high voltage. Guaranteed at slow line fluctuations or deviations of ±10%.

Alignment

Each lens is independently adjustable, mechanically and electrically, to permit optimum alignment. Lens current modulator can be switched on during critical centering of either condenser stigmator, objective stigmator or objective lens enabling the operator to use both hands during centering operation.

Maintenance

Swivel crane mechanism permitting rapid access to any part of optical system. Each pole piece easily removable for inspection or cleaning. Complete recentering of column is unnecessary when a single component has been re-inserted after cleaning. Unitized electronics, readily accessible, in separate cabinet. Five lens current stabilizing units interchangeable. Monitoring system for checking and measuring stabilities with built-in microvoltmeter.

Pumping System

Controlled by single, positive action, manually operated, valve. Rotary prevaccum pump (not in operating during actual use of the instrument). Buffer tank (reservoir). Mercury diffusion pump, water cooled. Oil diffusion pump, water cooled.

Evacuation Times

From cold start, 25 min. After changing filament, 5 min. After changing film camera, 10 min. After changing plate camera, 10 min. After changing specimen. 20 sec.

Water Cooling

For cooling of lenses and high vacuum pumps, 1 liter per minute required at 30p. s. i., 25 degrees C. Maximum temperature.

Power Consumption

4.5 K.V-a, cos $f = 0.7$.

Weight

Cabinet, approximately 1100 Ibs. Console, approximately 1550 Ibs.

Line Voltages

110-150, 200-250, 330-440 V. Permissible deviation ± 10% single phase 50 and 60 c/s. One of the great advances in modern systematic virology is the negative staining technic, which is showing rich rewards in the hands of Horne, Waterson, Wildy and others. The technic is relatively simple and involves spraying virus with 1 per cent phosphotungstic acid (PTA). It is by this technic that details of the intimate substructure of virus bodies are being worked out. For example, an adenovirus particle has been shown to be an icosahedron with 252 subunits each measuring 70Å.

Fascinating information about the nucleoprotein inner componenets of virus bodies is being obtained by examination of disrupted particles. A classic method of electron microscopic examination involves the making of a replica. This technic gives beautiful portrayals of virus crystals. Briefly, a film of virus crystals is coated with carbon, and the mount and crystals are dissolved. The mould or "replica" that remains is photographed. Such preparations illustrate the phenomenon of packing symmetry, and the polygonal shape of the constituent elementary bodies can be recognized. Another most important technic is that of ultramicrotomy, which enables very thin sections of cells or virus particles to be cut with a glass or diamond knife in an ultramicrotome.

Tissue culture cells or aggregates of virus elementary bodies are prepared for sectioning by fixing in osmium tetroxide and embedding in plastic. The use of this technic has revealed many significant features of virus development. For example, the inclusions of molluscum have been shown to be divided by septa; it has been found that the elementary bodies of herpes simplex form in the nueleus, and then pass into the cytoplasm obtaining a second outer membrane; and crystals of adenovirus have been found in the nucleus of HeLa cells, the individual elementary bodies forming a lattice, the center-to-center distance between the bodies being 60 to 65 mμ.

The vigorous study of the intricacies of virus substructure by means of various electron microscopic technics is likely to throw much light on the mode of virus multiplication, and is already providing a basis for a rational system of classification based on the conventional study of morphology.

FILTRATION

Older Types of Filters

The early work of Pasteur and others on the separation of bacteria from viruses was performed with porcelain and earthenware filter candles. Among the most extensively used has been the Pasteur-Chamberland that is available in many grades (L_1 to L_{13}) according to porosity. The Berkefeld and Mandler filters are made of Kieselguhr or diatomaceous earth, and have been used for preliminary clarification.

The disadvantages of these filters are that they are liable to become clogged, and can be cleaned only with difficulty, as by trypsin or heating to high temperature, when they may develop cracks or flaws. Another objection is that they absorb large amounts of virus.

The practical value of earthenware filters is thus limited to the recovery of relatively small amounts of virus from material contaminated with bacteria. The introduction of penicillin and streptomycin for destroying such contaminants seems to have diminished still further the usefulness of these filters in virus isolation work. Sales micro-porous filter are made of pure porcelain and do not contain diatomaceous earth. They are free of some of the disadvantages of the earlier types of candle, and can be used to prepare bacteria-free suspensions of virus containing material.

A good coarse filter for withholding bacteria is the Sitz asbestos disk type as used in general bacteriology, which is less subject to the disadvantages mentioned.

Elford's Collodion Membrane Filters

Thin membranes prepared either by Elford's method or by that of Bauer and Hughes can be manufactured in a range of porosities of known average pore diameter (APD) varying from 3μ to $10m\mu$. Parlodion or other nitrocellulose is first dissolved in a mixture of alcohol, ether, and acetone. Thereafter, to measured volumes of this solution are added varying concentrations of water and acetic acid; each mixture is poured into a flat bottomed glass cell, and is allowed to evaporate for a fixed time at constant temperature and relative humidity. By addition of water, the permeability of membranes can be increased, and by addition of acetic acid decreased. Membranes are cut into small disks, and the average pore size diameter is calculated by estimating the rate of water flow at known pressure for membranes of measured thickness, surface area, dry, and wet weight. Membranes are washed with water, sterilized by boiling or ultraviolet light, and stored in distilled water. For use, a membrane is clamped in a specially designed metal holder, and a positive or negative pressure of 25 cm, of mercury is applied. The size of viruses is estimated by testing for infectivity filtrates obtained from membranes of varying APD.

Estimates of the size of virus particles obtained. by Elford' method have often been larger than the figures obtained by the more precise methods of electron microscopy. Black, in reinvestigating this problem, has shown that the ratio of the size determined by electron microscopy to that obtained by filteration varies from 0.05-0.87. He concludes that, on the average, the figure for size determined by filtration should be multiplied by factor of 0.64, to give a figure close to that determined by electron microscopy of freeze dried particles.

Centrifugation

Although virus particles are very small, they nevertheless possess measurable mass and can be deposited by centrifugal force. The ordinary bucket type of horizontal centrifuge has served its purpose in the past, but even at 10,000 revolutions per minute (r.p.m.), the rate of sedimentation is slow, and several hours may be required to produce an appreciable deposition. Angle head centrifuges have been introduced as a means of reducing the distance to be traversed by sedimenting particles.

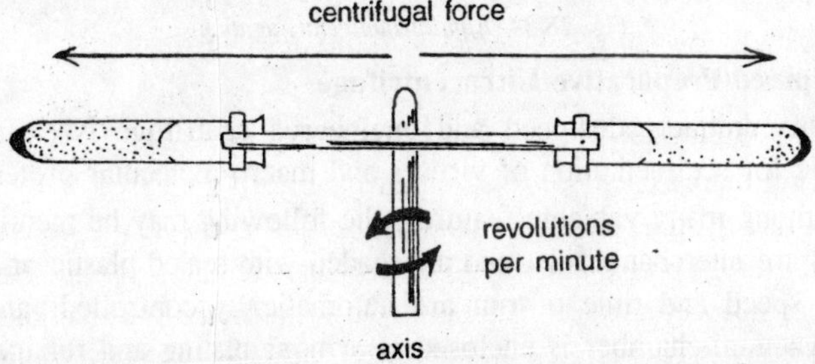

Fig. 18.33. Centrifugation. Extremely rapid rotation of liquids in centrifuge tubes produces a force in the tubes analogous to that of gravity, but much more powerful. Thus, suspended particles more dense than the liquid in a centrifuge tube separate (sink) rapidly. Following centrifugation, the supernatant fluid in the tube can be decanted (poured off), leaving the pellet of dense particles in the bottom of the tube.

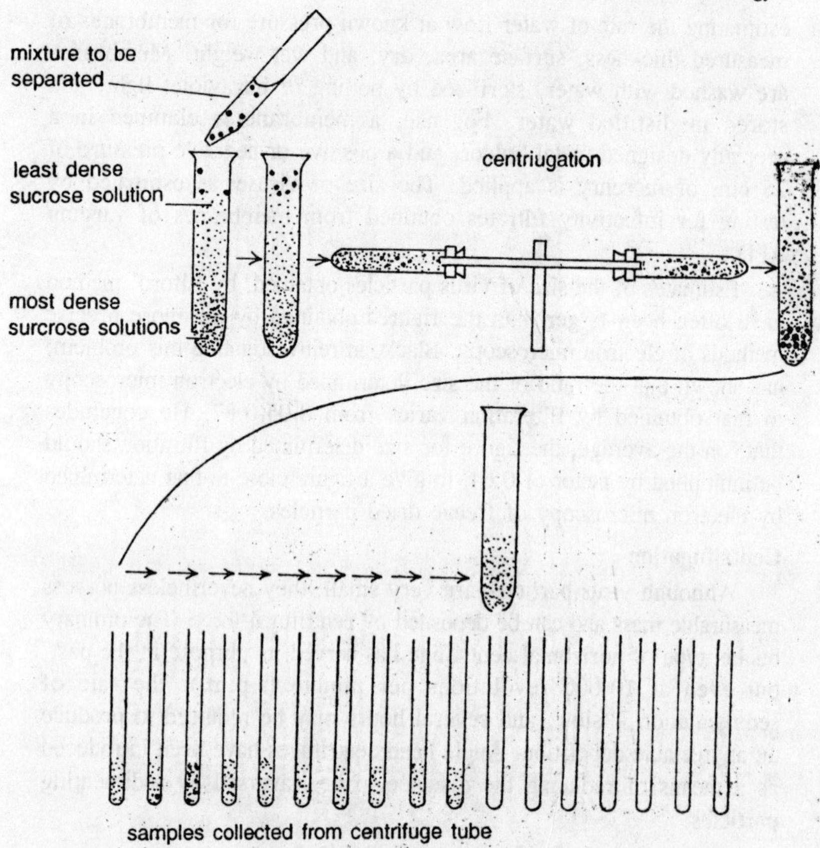

Fig. 18.34. Equilibrium centrifugation.

The Spinco Preparative Ultracentrifuge

This uniquely designed and engineered centrifuge is eminently suitable for sedimentation of viruses and macro-molecular proteins.

Among many valuable features, the following may be mentioned. Rotors are interchangeable, and are loaded with sealed plastic or metal tubes; speed and time of run are automatically controlled; and the rotor vacuum chamber is enclosed in armour plating and refrigerated so that low temperatures can be maintained during operation. Since the instrument is of the self balancing type, large scale preparative runs of quantities up to 1,600 ml. Of material may be handled without accurate balancing. Rotor #21, with 10 tubes, inclined at an angle of 18°, has a total capacity of 940 ml. A maximum centrifugal force of 59,000 g. At 21,000 r.p.m. Is attainable in an acceleration time of 15

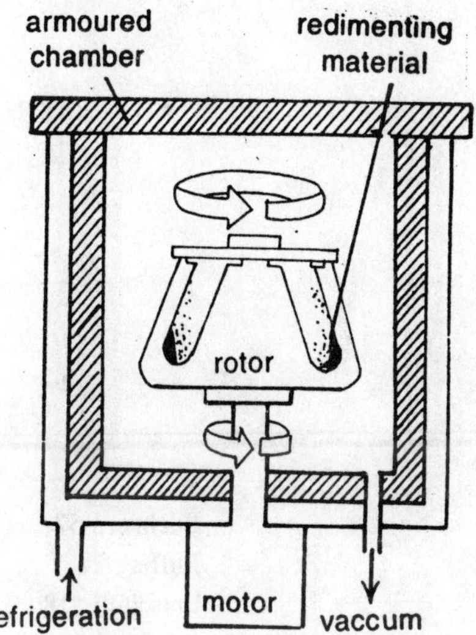

Fig. 18.35. The preparative ultracentrifuge. The sample (i.e., homogenate) is contained in tubes that are inserted into a ring of cylindrical holes in a metal rotor. Rapid rotation of the rotor generates immense centrifugal forces, which cause particles in sample to sediment. The vacuum tends to reduce friction, preventing heating of the rotor and allowing the refrigeration system to maintain the sample at 4°C.

minutes. Rotor #40, with 12 tubes, inclined at an angle of 26°, has a capacity of 162 ml.; a maximum centrifugal, force of 144,700 g. At 40,000 r.p.m. is attainable in an acceleration time of 5 minutes.

INDEX